Clay Water Diagenesis During Burial: How Mud Becomes Gneiss

Charles E. Weaver
and
Kevin C. Beck

School of Geophysical Sciences, Georgia Institute of Technology, Atlanta, Georgia

Charles O. Pollard, Jr.

School of Geophysical Sciences, Georgia Institute of Technology, Atlanta, Georgia

Appendix: Semidisplacive Mechanism for Diagenetic Alteration of Montmorillonite Layers to Illite Layers

THE
GEOLOGICAL SOCIETY
OF AMERICA

SPECIAL PAPER 134

Library of Congress Catalog Card Number: 74-153907
I.S.B.N. 0-8137-2134-2

Published by
THE GEOLOGICAL SOCIETY OF AMERICA, INC.
Colorado Building
P.O. Box 1719
Boulder, Colorado 80302

Printed in the United States of America

The printing of this volume has been made possible
through the bequest of
Richard Alexander Fullerton Penrose, Jr.

Acknowledgments

This study was supported by research grants from the United States Department of the Interior, Office of Water Resources Research, as authorized under the Water Resources Act of 1964, Grant A-008-6-A and the National Science Foundation, Grant GA-1330. The authors are indebted to G. B. Baker, M. G. Fey, and H. L. Ellenwood of the Chevron Oil Company and A. E. Worthington of the Standard Oil Company of California for supplying samples and information. The cooperation of the Shell Oil Company is appreciated.

Contents

FIGURES

Appendix

Abstract

Mineral and chemical studies of muds samples (sidewall cores) from 4233 ft to 16,450 ft deep in the Gulf of Mexico (courtesy Chevron Oil Company) and from the surface to 24,003 ft in the Anadarko Basin, Oklahoma (courtesy Shell Oil Company) were studied in order to obtain information on the diagenesis of muds during deep burial.

With depth the montmorillonite in these montmorillonite-rich shales is altered to mixed-layer illite-chlorite-montmorillonite; the regularity of the mixed-layer phase tends to increase with depth. Much of the interlayer hydroxy material is Al and Fe, acquired before deposition. K is acquired from the river and seawater, and after deposition, from K-feldspar. Increased lattice charge in the montmorillonitic layers (largely beidellite) is due to the reduction of iron in the octahedral layer and the incorporation of additional Al in the tetrahedral layer. It is suggested that the latter phenomenon is caused by the migration of interlayer Al into the hexagonal holes in the oxygen sheet. Most of the loss of expandable layers occurs by 12,000 ft.

As depth and temperature increase, kaolinite is destroyed, and the Al is deposited between the expanded layers as hydroxy-Al to produce layers of dioctahedral chlorite. Both discrete dioctahedral chlorite and mixed-layer illite-dioctahedral chlorite-montmorillonite (ultimately mixed-layer illite-chlorite) are formed. During regional metamorphism the sequence is kaolinite and mixed-layer illite-montmorillonite $\longrightarrow$ mixed-layer illite-dioctahedral chlorite $\xrightarrow{K, Mg, Fe}$ muscovite + chlorite.

The Al_2O_3 content of the bulk samples ranges from 9.94 to 17.47 percent. This is equivalent to approximately 55 to 75 percent clay minerals. The Al_2O_3/K_2O, Al_2O_3/MgO, and Al_2O_3/Fe_2O_3 values of the Chevron and Mississippian (except for Al_2O_3/MgO) samples indicate a deficiency of K, Mg, and Fe with respect to the Paleozoic and Precambrian shales. There is an indication that these cations increase with depth and may have migrated upward from the interval where granitization is taking place. These montmorillonite-rich clays cannot be converted to the typical illite-chlorite clay suite of the older shales without the addition of K, Mg, and Fe from external sources. Dioctahedral chlorite, both as discrete chlorite and as mixed-layer illite-chlorite, is present in most Paleozoic and Precambrian shales and will be even more abundant in deeply buried Tertiary shales unless cations are added from outside the system. The data indicate that, in areas where the conversion of montmorillonite-rich clays to illite-chlorite clays occurs, the geothermal gradient is relatively high, and K, Mg, and Fe are added from below. The possibility exists that geothermal gradients were higher in the Paleozoic and Precambrian, particulary preceding the middle Carboniferous break-up of the continents.

The total exchange cations (Na + K + Mg + Ca) average 30 meq/100 g down to approximately 8000 ft; deeper samples average 20 meq/100 g. The C.E.C./Al_2O_3 values decrease to 10,000 ft, then remain constant, confirming the X-ray interpretation. The more mobile Na replaces Mg to become the dominant exchangeable cation in the shallow samples. As the expandable layers are converted to nonexpanded layers, the weakly bonded Na is released to the pores, and Ca is selectively retained. Most of the exchangeable Mg is used to form dolomite or is flushed into the sands where it combines with montmorillonite to make chlorite.

The pore water content decrease to 10,300 ft abruptly increases, as high pressures are encountered, and then systematically decreases. The cation concentration in these waters is two to three times that of seawater. The cation concentration increases to 10,000 ft, abruptly decreases by 20 percent, then remains relatively constant. Na and K are more concentrated than seawater; Ca and Mg are less concentrated. K and Mg increase with depth; this is presumably a function of the increase in temperature. In the shallow samples the Na concentration increases as the pore water decreases. In deeper samples the ratio remains relatively constant.

The anion concentration is $HCO_3 > SO_4 > Cl$. Cl systematically decreases with depth and is one-fourth the concentration of seawater by 10,000 ft. Presumably, this is due to selective flushing. Over the same depth interval, SO_4 increases by a factor of 6 to 7, comparable to the decrease in pore water, suggesting concentration by selective filtering. HCO_3 increases in concentration to 10,000 ft and then remains constant. The high HCO_3 values are due to the decomposition of organic matter and calcite. Functional organic groups are released from the clay minerals as the temperature and Na concentration increase.

During burial a physical permeability barrier is formed by the rapid dewatering of the top of the thick water-rich mud section. Upward migrating Ca precipitates as calcite increasing the effectiveness of the buried mud. Na diffuses through the barrier more easily than water; Cl diffuses through more easily than SO_4 and HCO_3.

Introduction

It has been noted by a number of people that montmorillonite-rich clays tend to lose their ability to expand with depth. It has been concluded that montmorillonite alters to illite with depth. Little data, other than X-ray analysis, is available. The nature of the clay minerals, chemical changes, and mechanisms responsible for the changes are speculative.

There is little or no data on the chemistry of the interstitial water and the exchangeable cation in deeply buried clays. The relation between the pore-water chemistry of muds and sandstones is inferred almost entirely from sandstone data.

The mechanism of clay compaction, dewatering, and fluid pressure is not well understood.

In order to obtain additional data on these problems and hopefully some answers, a series of sidewall cores of muds and silty muds was obtained from a deep well in the South Pass Area of Louisiana. The cores and information about the well were kindly supplied by the Chevron Oil Company. Supplemental information from Gulf Coast and Paleozoic sediments is based on data acquired by the senior author while employed by the Shell Oil Company.

The samples from the Chevron well range in depth from 4233 ft to 16,450 ft. The top of the Pliocene is at approximately 4000 ft and the top of the Miocene approximately 9000 ft. The well bottomed in the upper Miocene. Chevron paleontologists report that the depositional environments systematically increase from inner neritic at 4233 ft to outer neritic at 9100 ft. All deeper samples indicate a bathyal environment. There is a marked increase in numbers of tests of deep-water benthonics from 10,930 ft to total depth. Abnormally high fluid pressures were encountered at 10,230 ft and continued to total depth.

Sample Condition and Analytical Techniques

SAMPLES

The samples as received were 1 to 2 in. long and 1 in. in diameter. They were sealed in glass jars and were moist. Even the deepest samples were somewhat pliable and resembled a dense mud more than a shale. A tissue-thin covering of drilling mud coated much of the surface of the cores.

There was a major concern about contamination of the samples, particularly the pore waters, by drilling fluids. However, the consistency of the data and the regularity of the variations suggest that contamination is minor.

Two sidewall cores from another well in the area (courtesy of Gulf Oil Company) were analyzed and gave similar results for the composition of the interstitial water. In the deeper portion of the Chevron well, water flowed from the formation to the well. If fluid had flowed from the mud to the formation a more uniform ion suite would be expected. The *p*H of the mud ranged from 10 to 11, whereas that of the samples was 8.

Some of the data could be suspect, but in view of the scarcity of data on the interstitial water of deeply buried muds, we feel it should be presented.

ANALYSES OF SOLIDS

Chemical analysis of the solids was carried out by atomic absorption spectrometry. Fe^{2+} was determined by the volumetric method of Wilson (1960). Mineral identification and quantitative estimations were based on X-ray diffraction analysis. Standard techniques were used (Weaver, 1967, p. 44–50).

ANALYSES OF EXCHANGE IONS

The exchange cations Na^+, K^+, Mg^{2+}, and Ca^{2+} were determined on samples leached of interstitial waters by replacement with Ba^{2+}, using an O.1 *N* solution of $BaCl_2$ adjusted to *p*H 7 with $Ba(OH)_2$. Cation analysis was made by atomic absorption spectrometry.

ANALYSES OF INTERSTITIAL IONS

Samples of known weight (air dried and 110° C basis) were leached with distilled water, and the water was analyzed. This method is open to errors in respect of *p*H, HCO_3, and ions associated with soluble minerals. The relative proportions of ions thus obtained were converted to absolute concentrations from estimates of the amount of interstitial water.

The major cations were analyzed by atomic absorption spectrometry. Chloride was determined coulometrically (Buchler-Cotlove chloridometer) and checked by titration with mercuric nitrate in alcoholic medium using diphenylcarbazone and 2-nitroso-1-naphthol as indicator buffer. Sulphate was determined spectrophotometrically at 330 mμ using barium chloranilate (Klipp and Barney, 1959). Interfering cations were removed by ion exchange. Bicarbonate was determined by titration with standard acid (Rainwater and Thatcher, 1960); *p*H was determined potentiometrically. As the *p*H of the leach waters was greater than 8.2, significant $CO_3^{=}$ is present, but in view of uncertainties due to small sample and the somewhat doubtful significance of the leaching procedure, total C is reported as HCO_3.

Mineralogy

NON-CLAY MINERALS

Quartz is present in all samples. X-ray patterns of bulk samples suggest quartz is most abundant in samples above 8000 or 9000 ft and below 15,000 ft. The Al_2O_3 distribution is roughly the inverse of this. The resistivity values obtained from the electric log show a similar pattern, that is, high resistivity values for the shallow and deep mudstones and lower values for the interval from 10,400 ft to 15,000 ft.

Minor amounts of Na-feldspar (3.20 Å) and K-feldspar (3.24 Å) are present in all samples; the former generally is the more abundant. K-feldspar appears to be less abundant in the deeper samples. Dolomite in small amounts (< 5 percent) is detectable in all but one sample. Small amounts of calcite were found in three samples. Minor amounts of siderite and pyrite are present in samples deeper than 15,500 ft. Trace amounts may occur in some shallower samples. Rounded flakes of muscovite and biotite are present in the silt fraction.

Barite was detected in most samples and is particularly abundant in the thin layer of drilling mud coating the samples. Because of the relatively high SO_4 content of the interstitial waters, it is possible that some of the barite is authigenic, but most probably comes from the drilling mud.

CLAY MINERALOGY

Pliocene-Miocene

The vertical distribution of minerals in the Chevron well is similar to that previously reported from the Gulf Coast area (Burst, 1958; Powers, 1958; Weaver, 1958b), though there are significant differences in interpretation. The shallow muds are composed largely of montmorillonite with lesser amounts of illite, chlorite, and kaolinite. The suite is similar to the one that is presently carried by the Mississippi River. The deeper samples are composed largely of mixed-layer illite-chlorite-montmorillonite with smaller amounts of illite, kaolinite, and chlorite. Figure 1 shows the vertical distribution of the clay

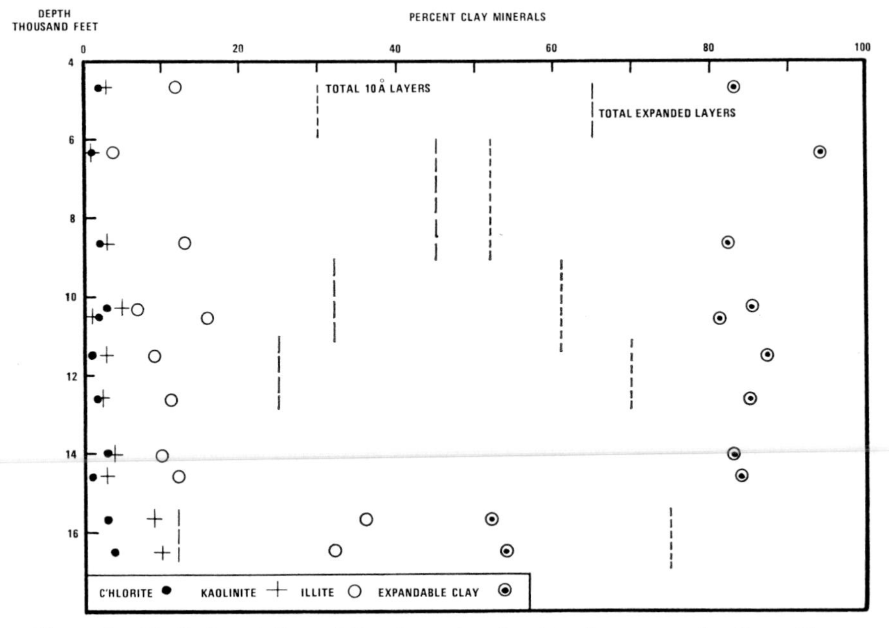

Figure 1. Vertical variation of the clay mineral suite of core samples from Chevron well.

minerals. The absolute percentage values are not considered to be too significant, but the relative differences are reasonably reliable.

Kaolinite is more abundant than chlorite. Both are present in amounts less than 5 percent down to 14,600 ft; deeper samples contain significantly larger amounts of kaolinite. The kaolinite/chlorite ratio appears to increase slightly with depth.

Two Chevron samples (4233 and 15650) were treated according to the method of Alexiades and Jackson (1966) as follows: 525° C for 4 hrs, boil 2.5 min in 0.5 N NaOH to determine the amount of kaolinite in the bulk samples. The shallow sample lost sufficient Al to account for 3.3 percent kaolinite; the deep sample enough to form 7.1 percent kaolinite. These figures are in good agreement with the estimates based on X-ray data. The silica dissolved by this method (6.5 percent and 8.6 percent SiO_2, respectively) is more than required to form the amount of kaolinite indicated by the Al_2O_3 values and presumably indicates the presence of amorphous silica.

Illite remains relatively constant, approximately 10 percent, to near 15,000 ft where it increases to 30 percent. The amount of expanded clay in the clay fraction (montmorillonite and mixed-layer clay) is approximately 85 percent to 15,000 ft where it decreases to 55 percent. X-ray patterns of glycolated samples have a relatively sharp 17 Å peak down to 10,530 ft; however, even at 4233 ft the third-order 5.66 Å peak is quite broad, extending toward 5.0 Å. The third-order peak indicates there are some contracted 10 Å layers (on the order of 10 to 20 percent) interstratified with the montmorillonite layers even in the shallowest samples. This mixed-layer effect in the shallow material is probably largely inherited and reflects the original detrital composition of the expanded clay.

Calculations (MacEwan and others, 1961) show that for randomly interstratified 10 Å and 17.5 Å layers of limited thickness there is little shift in the position of the 001, 17.5 Å maxima until the 10 Å layers comprise about 60 percent of the total. As the proportion of 10 Å layers increases further the 001/001 maxima broadens and decreases in intensity as the proportion of 17.5 Å layers decreases. Peak broadening also increases with decreasing thickness of the crystallites. The 002/003 (5.0 Å/5.7 Å) peak shifts position in a regular fashion as the proportion of the two components varies and thus is a more sensitive measure of mixed-layering. Reynolds (1967) has shown that in order to obtain 001/001 peaks intermediate in position between 10 Å and 17 Å it is necessary that the mixing not be completely random, but show some degree of order. Presumably, peak sharpness increases with increased order.

The 4233-ft sample has a relatively discrete 5.7 Å peak; samples down to a depth of 8650 ft have a fairly discrete 002/003 peak with a broad top extending from approximately 5.4 Å (6:4 illite-montmorillonite) to 5.7 Å (montmorillonite). From 9100 to 11,020 ft the peak ranges from 5.0 Å to 5.7 Å and is flat topped. In the latter samples mixing must be quite random, and the illite layers comprise approximately 50 to 60 percent of the mixed-layer phase.

Samples below 11,020 ft, with one exception, do not have discrete 17 Å peaks. The broad 002/003 peak extends from 5.0 to 5.7 Å. The low-angstrom portion of the peak is more intense than the high-angstrom portion with the relative difference generally increasing with depth (Fig. 2). The relative intensity of the peak also decreases with depth. A relatively discrete 5.2 Å peak (7:3 illite-montmorillonite) can be seen in the pattern of the 12,020-ft sample. The proportion of 10 Å layers in the mixed-layer clay appears to continue increasing slightly in the deeper samples. The deepest samples have ratios in the 7:3 to 8:2 range.

It is difficult to determine if the abrupt increase in the illite content of the deeper samples is real or is due to the fact that the mixed-layer peaks (001/001 and 001/002) are very close to 10 Å. The former appears to be the more probable.

The 11,520-ft and deeper samples have broad scatter from the 001/001 (10 Å/17 Å) reflection ranging from 12 to 20 Å. Low-angle peaks occur ranging from approximately 27 to 37 Å. The presence of the 27–37 Å peaks indicates that there is some regularity to the interlayering.

The pattern for the 12,020-ft sample has a discrete peak at 27 Å and a broad second-order peak at approximately 13.5 Å. These values indicate the mixed-layer clay contains a significant number of regularly alternating 10 Å and 17 Å layers.

With increasing depth the intensity of the low-angle peak increases systematically relative to that of the 001/001 peak. At the same time the peak shifts toward 37 Å. The 001/001 peak also shifts and narrows. At 12,890 ft it ranges from 12 Å to 21 Å; at 14,010 ft it ranges from 14 Å to 19 Å; at 14,538 ft it ranges from 17 Å to 22 Å. Both sets of peaks indicate an increase in the proportion of 10 Å layers and an increasing tendency for 10 Å layers to alternate regularly with a 17 Å layer.

The 15,000-ft and 16,000-ft samples have a broad 001/001 peak ranging from 10 to 17 Å with a weak maximum near 14 Å to 15 Å. The two 16,000-ft samples have fairly discrete breaks in the slope of the background curves at 2θ values equivalent to approximately 28 Å. This spacing, along with the 14 Å to 15 Å spacing, suggests that regularity has increased though there is little if any change in the proportion of the two types of layers.

Figure 1 shows the estimated amount of total 10 Å layers based on the above evaluation. The amount of 10 Å layers, of the total clay suite, increases from a minimum of 20 to 30 percent in the shallowest samples to approximately 75 percent in the deeper samples.

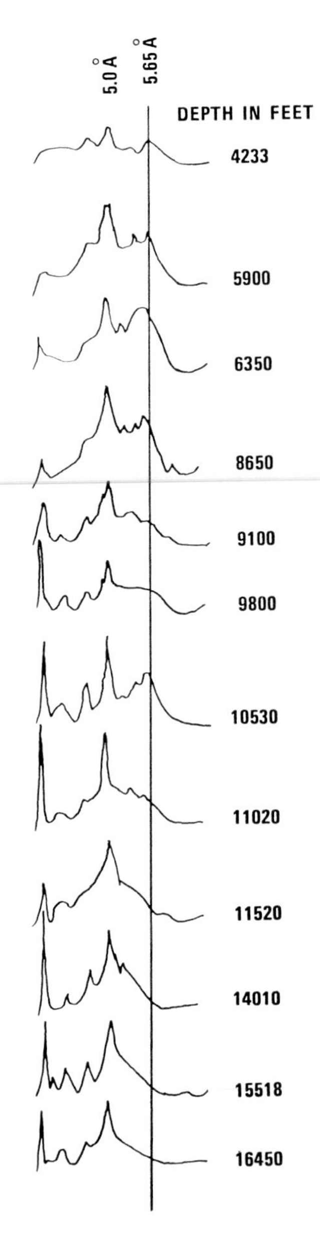

Figure 2. X-ray diffraction patterns of oriented Chevron samples. The 5.65 Å line marks the position of the 003 montmorillonite reflection. The 5 Å peak is the 002 illite reflection.

Most of the contraction of montmorillonite layers to 10 Å occurs between 4000 and approximately 12,000 ft. It is difficult to determine what is happening below 10,000 to 11,000 ft in the Chevron well. From 10,000 to 11,000 ft to 16,450 ft it appears that there is some increase in the proportion of 10 Å layers, but the major change appears to be an increase in regularity. The C.E.C./Al_2O_3 values indicate there is little change in the proportion of expanded layers below 11,000 ft (Fig. 12).

Miocene samples from other wells in the Gulf Coast area indicate the next diagenetic stage is the development of a mixed-layer phase which affords a relatively sharp 13 to 13.5 Å peak (also ~ 30 Å, 9.3 Å to 9.4 Å and 5.2 Å to 5.3 Å peaks). The 13 Å phase is quite stable and once formed persists for at least an additional 5000 ft of depth (Shell-Continental, Federal Block A-72). This means that such a phase would exist to a depth of at least 23,000 ft in the area of the Chevron well. Chlorite and illite appear to increase slightly, and kaolinite decreases during this stage.

It should be pointed out that most estimates, including the above, of the amount of 10 Å layers in mixed-layer clays, are based on assuming the glycolated clay consists of a mixture of 10 Å and 17 Å layers. X-ray analyses of heated samples (300° C to 500° C for 2 hours) indicate most of the mixed-layer clays in this study contain a 14 Å chloritic component. Thus the estimated amount of 10 Å layers is high. The presence of chloritic layers is indicated by the low-angle asymmetry of the 10 Å peak and the high-angle asymmetry of the 5 Å peak. In the Chevron samples there does not appear to be any systematic increase in or decrease in the amount of chloritic material with depth.

Theoretical scattering distribution curves were calculated for a 1:1 random mixture of 14 Å chlorite layers and 17 Å montmorillonite layers. The relative intensities of the first four reflections are:

Peak	Relative Intensity
001/001	100 sharp
002/003/002	3 broad
004/003	5 relatively sharp
005/004	16 sharp

Thus, it is relatively difficult to detect the presence of mixed-layer chlorite-montmorillonite in the presence of other clay minerals. The 004/003 peak would be relatively more intense when the chlorite is dioctahedral.

An attempt was made to obtain some information on the nature of the interlayer material (Table 1) by boiling some Chevron samples for 5 minutes in NH_4F (Rich and Obenshain, 1955).

TABLE 1. INTERLAYER MATERIAL

Depth Feet	Percent Al_2O_3	Al_2O_3/ Al_2O_3%*	Percent Fe_2O_3	Fe_2O_3/ Al_2O_3*	Percent MgO	MgO/ Al_2O_3*
4233	2.8	0.28	3.7	0.37	0.31	0.031
9100	3.2	0.25	6.5	0.51	0.43	0.034
12020	4.2	0.24	3.4	0.19	0.28	0.016
15044	3.2	0.19	3.9	0.23	0.40	0.024

* (bulk)

This technique presumably dissolves amorphous material, siderite, and a portion of the talc layer as well as the hydroxy interlayers: however, the data does suggest that the interlayers are probably dioctahedral. After the treated samples were heated to 300° C, the 10 Å peak was considerably sharper than before, confirming the presence of chlorite layers in the mixed-layer phase of the untreated samples. The amount of dissolved material per amount of Al_2O_3 (used as a measure of the clay content of the bulk samples) is less in the deeper samples and the 10 Å peaks are broader which could indicate the interlayer material in the deeper samples is better organized and less easily extracted.

On the basis of the K_2O and Al_2O_3 values, the deeper samples (below 10,000 to 11,000 ft) should have fewer 10 Å layers than indicated by the X-ray analysis. In part this is due to the presence of the chlorite layers in the mixed-layer phase. The chloritic layers would make up 30 percent, assuming 10 percent K_2O in the illitic layers.

Samples 11,020 ft and shallower were soaked in a 1N KOH solution overnight. X-ray patterns of the samples, dried at room temperature, had relatively flat-topped peaks extending from 10 Å to 11 Å in the shallower samples and 10 Å to 13 Å in the deeper samples. Potassium-saturated samples were heated at 200° C for three days and treated with ethylene glycol. As with the unheated samples, a larger proportion of layers contracted in the shallower samples than in the deeper samples.

These data indicate the expanded layers vary in their ability to contract and that the shallower samples have a higher proportion of easily contracted layers. Presumably, in the deeper samples, the easily contracted layers have been contracted under natural conditions.

Mississippian

The Springer Shale of Upper Mississippian age of Oklahoma is similar in nearly all aspects to the Gulf Coast Tertiary muds, but illustrates a more advanced stage of diagenesis. Montmorillonite is converted to a mixed-layer illite-chlorite-montmorillonite with depth. The expanded clays vary from 50 to 60 percent of the clay suite; illite varies from 10 to 25 percent; the general suggestion is that the deeper shales contain more illite. The sandstones in both areas commonly contain an abundance of poorly crystallized chlorite.

Chlorite most commonly ranges from 10 to 15 percent and kaolinite, from 5 to 20 percent. Analyses of samples from approximately 50 wells indicate that the abundance ratio of these two minerals does not vary regularly with depth; however, in the deepest section studied, 22,000 ft to 24,000 ft, the chlorite/kaolinite ratio systematically increased. Figure 3 is a 65-mi cross section showing the variations in clay mineral suite of the Springer Shale at depths ranging from 6500 to 24,000 ft. There is little vertical change within the Springer of the individual wells, but laterally and with increasing depth kaolinite and mixed-layer illite-chlorite-montmorillonite decrease and illite and chlorite increase. Most of the increase occurs in the deepest section. The chlorite plus kaolinite content of the Springer Shale is considerably larger than in the Chevron muds.

The 5:4.75 Å peak height ratio of the untreated Springer Shale samples increases from well to well as sections become deeper. This indicates that the illitic component (which contains some chlorite layers) is increasing more rapidly than the discrete chlorite. This is also true of the Chevron samples.

The change in the expanded layers with depth is similar for the Springer and Pliocene-Miocene montmorillonites, though the latter samples have not progressed quite as far. In the shallower Springer samples, mixed layering is indicated by high background in the low-angle region and broad 14 Å to 17+ Å peaks of the glycolated samples. As depth increases, this peak shifts to a broad peak ranging from 10 Å to 14 Å. The broadness of these peaks indicates

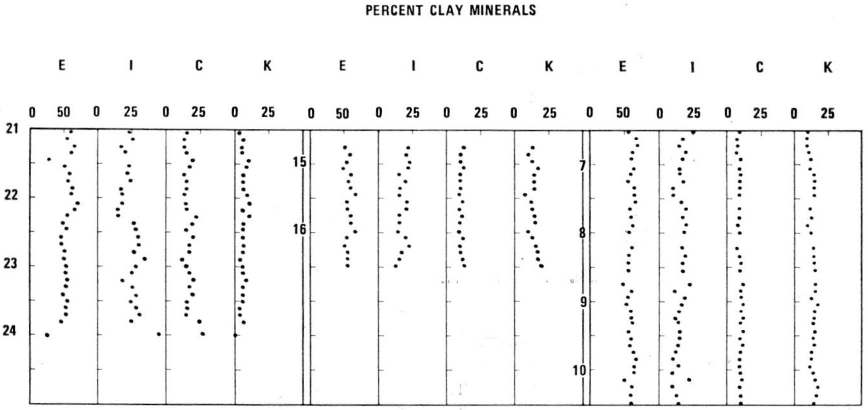

Figure 3. Composition of clay mineral suite of Springer Shale samples from three wells in the Anadarko Basin, Oklahoma. On line of section the middle well is 25 mi from left well and 40 mi from right well. E, expandable clay (montmorillonite and mixed-layer illite-chlorite-montmorillonite); I, illite; C, chlorite; K, kaolinite. Vertical scale is in thousands of feet.

that the mixed layering is fairly random. With increasing depth a discrete 13.0 Å to 13.5 Å mixed-layer peak develops accompanied by a series of peaks at 30 Å to 32 Å, ~ 9.5 Å, 5.2 Å to 5.3 Å and 4.8 Å to 4.9 Å. The relative sharpness and intensity (compared to the 10 Å peak) of this peak increases to a considerable depth, then the 10 Å increases at the expense of the 13 Å peak. The 13 Å mixed-layer phase first occurs at a depth of 14,000 to 15,000 ft and persists to a depth of at least 23,820 ft.

A core from 24,000 to 24,003 ft contains primarily the clay minerals chlorite and illite. Unfortunately at this depth a different formation is encountered, and it is not certain that the clay suite represents a continuous diagenetic sequence. The chlorite has an 003 peak (4.75 Å) with four times the intensity of the 001 peak (14.2 Å). The 003 peak is the most intense chlorite peak. This distribution of peak intensities is characteristic of a dioctahedral chlorite (Fig. 4).

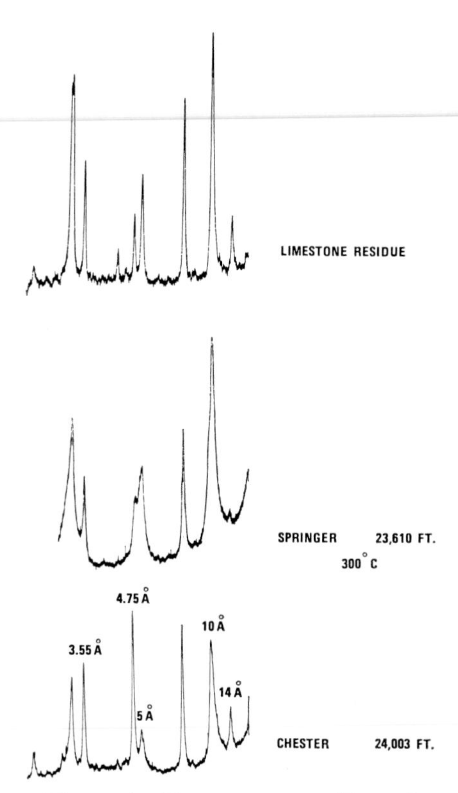

Figure 4. X-ray patterns illustrating variations in the relative intensity of the 00ℓ reflections of trioctahedral and dioctahedral chlorite. Upper sample contains dioctahedral chlorite and relatively pure illite. The middle pattern shows the intensities when both types of chlorite are present, and the bottom pattern, when the chlorite is predominantly trioctahedral.

Table 2 gives the peak positions of the 13 Å phase (concentrated) after various treatments.

The low-angle peak is fairly well developed indicating some regularity; however, the 00ℓ sequence is not regular. When the Springer samples are heated at 300° C for two hours the 001/001 peak ranges from 10 Å to 12 Å. When the same sample is reheated at 300° C for several hours, the peak ranges from 10.0 Å to 10.5 Å with a shoulder at 12 Å. As with hydroxy-polymers formed in the laboratory (Rich, 1968), many of those in the mixed-layer phase of the Springer clays are not well developed and are relatively unstable.

Some indication of what happens to the 13 Å phase with increasing depth of burial can be obtained from the X-ray patterns of the heated (300° C) samples. A plot of the peak width at half-height of the 10 Å to 11 Å peak (Fig. 5) shows there is

TABLE 2. DIFFRACTION MAXIMA OF 13 Å MIXED-LAYER CLAY

Treatment	Layer Thickness	Å	Å	Å	Å
Mg	10/14	28	12.4	9.3?	4.75 – 5.0 4.88?
Ethylene Glycol	10/14/17	31.5	13 – 14.5	9.3 – 10.0	5.25 + 4.75 –5.0?
300° C	10/14	25	10 – 10.5 11 – 12	9.0	4.75 – 5.0 4.93?

a fairly abrupt decrease in peak width at 22,300 ft. Above this depth, the apex of the peak ranges from 10.0 Å to 10.5 Å and the base of the peak flares out to 14 Å; deeper samples usually have a narrower 10 Å peak and small discrete peaks at 12 Å and 8 Å. The latter two peaks indicate the presence of regularly alternating 14 Å chlorite layers and 10 Å layers. The 7 Å peak of the shallower samples tends to be more asymmetrical toward the low-angle side (10 Å) than the deeper samples. In many of the heated samples both the 14 Å and 3.5 Å peaks are more asymmetrical in the direction of the 10 Å series of 00ℓ peaks after heating than before, suggesting the presence of some montmorillonitic layers intergrown with the chlorite layers. Due to interfering peaks it is difficult to tell how this asymmetry varies with depth (it is easier to observe in the deeper samples). Most of these heated samples have minor peaks in the 15 Å to 17 Å, 7.5 Å to 9.5 Å and 5.6 Å to 6.0 Å range. Presumably these peaks are due to a separate mixed-layer phase relatively rich in chlorite. The decrease in peak width at 22,000 ft is accompanied by a decrease in the mixed-layer clay and kaolinite and an increase in the amount of discrete illite and chlorite.

Another abrupt decrease in peak width occurs between 23,820 and 24,003 ft. In the deepest sample available (24,003 ft) mixed-layer clay comprises approximately 25 percent of the clay suite. The glycolated mixed-layer peak occurs as a shoulder on the 10 Å peak extending to 12 Å. Thus, this mixed-layer phase proba-

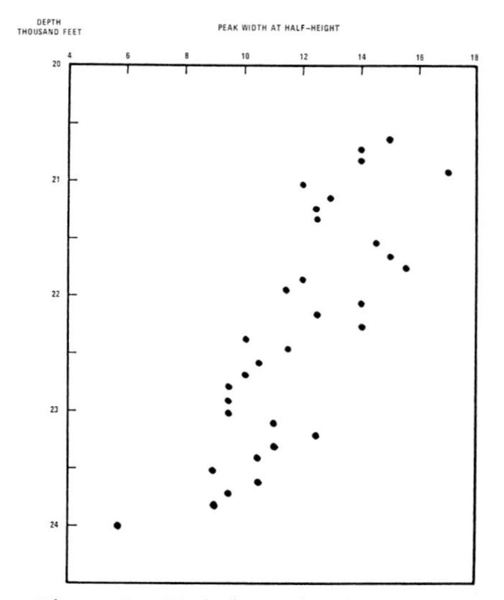

Figure 5. Variation with depth of the half-height width of the 10 Å X-ray peak. Samples heated to 300° C. Springer Shale samples from Anadarko Basin, Oklahoma.

bly contains a higher proportion of 10 Å layers than the 13 Å phase. At 300° C a small peak develops at 11.5 Å indicating some chloritic layers are present in the mixed-layer phase and the 10 Å peak increases in intensity showing the presence of some montmorillonitic layers. With the disappearance of the 13 Å phase (55 percent of the clay suite), the discrete illite (10 Å peak) content increases by 15 percent and the chlorite by 10 percent.

It would appear that during the final stages of diagenesis of the mixed-layer phase, illite and chlorite layers tend to form packets of sufficient thickness to diffract as discrete units; however, unmixing is not usually carried to completeness and varying amounts of chlorite layers occur intergrown with the illite layers.

The results of the heat treatments indicate the 13 Å phase contains an appreciable number of chlorite layers. The peak shift caused by glycolation indicates considerable montmorillonite is present. Thus a reasonable estimate would be that the three types of layers are present in approximately equal proportions.

The chlorite in the deepest sample is at least in part dioctahedral and presumably much of the diagenetically formed chlorite layers in the shallower samples and in the Chevron samples are dioctahedral (this is confirmed by the chemical data). During more advanced stages of diagenesis Al from kaolinite must be deposited as hydroxy-Al interlayers in montmorillonite. Minor amounts of cristobalite are present in the deep Springer Shale. This may be residue from the dissolved kaolinite.

The dioctahedral aspects of the newly formed chlorite can best be seen in the relative intensities of the 00ℓ peaks, even though there is some interference from the mixed-layer phase (Fig. 4). The 003/001 values for the detrital chlorites are similar to one; the deep Springer chlorite values range from 2.5 to 5 and increase with increasing depth. The values for the deep Springer samples are from a mixture of detrital trioctahedral chlorite and diagenetic dioctahedral chlorite. These two chlorites are present in approximately equal amounts.

It should be pointed out that chlorite with these relatively high 003/001 values is rare in the stratigraphic record; however, it is possible that mixtures of this diagenetic chlorite (003/001 $\sim$ 4) and detrital chlorite (003/001 $\sim$ 1) might be easily overlooked.

The deepest samples examined in this study and in other studies, indicate montmorillonitic material is not completely converted to mica-like layers under normal conditions, that is, 25,000 ft burial, less than 200° C. Much of the loss of expanded layers is due to the formation of chlorite layers rather than mica. Some expandable montmorillonite layers can persist until a relatively high grade of metamorphism is reached.

Mechanism

Much of the chloritic interlayer material in the shallow samples is apparently acquired while the clay is in contact with the seawater and is trioctahedral (Grim and Johns, 1954). Both Mg and Fe are probably present with most of the Fe being acquired before the clay reaches the open sea.

Analyses of samples from rivers draining into the Gulf of Mexico indicate much of this interlayer material is already present, presumably inherited from older sediments or newly formed in soils.

On the basis of soil studies (Glenn and Nash, 1964; Post and White, 1967; Glenn and others, 1960) in the Mississippi drainage basin, it seems quite likely that the detrital montmorillonitic clays have a considerable amount of hydroxy-Al in the interlayer position.

Little additional hydroxy-Mg is acquired during shallow burial (see discussion on interstitial water), but some Fe from free iron oxides may move into the interlayer position. Some discrete chlorite may be formed under these shallow conditions, but the amount is small. The shallow Springer and Pliocene-Miocene clays usually contain less than 10 percent chlorite, usually half this amount, and much of this is detrital.

Some K is acquired while the clay is in contact with the river and seawater and higher charged layers contract to 10 Å (Weaver, 1958a, p. 856–859). The number of layers that would contract under these conditions would vary with the composition of the montmorillonite, but would be relatively minor in the samples cited. Additional K is present as exchangable cations, in the interstitial water, in organic material, and in K-feldspar. However, the total K_2O content is insufficient to allow for the complete conversion of the montmorillonite to 10 Å mica-like layers.

Some rearrangement of the K takes place at shallow depth. Mud wafers of a K-montmorillonite and a Ca-stripped illite were placed in contact and kept moist. Within a few weeks sufficient exchange had taken place so that a considerable portion of the "expanded illite" layers had contracted to 10 Å. A mud was made composed of Ca-stripped illite, K-feldspar, and distilled water. After several weeks there was a significant increase in the amount of 10 Å material.

This indicates that under the conditions of shallow burial, K is relatively mobile, and sufficient K should be able to migrate into the highly charged layers and cause contraction. Treatments with KOH and drying of samples at 200° C indicate that some potentially contractable layers (under these relatively drastic conditions) are present down to 6300 ft. Deeper samples do not appear to show additional contraction under these conditions.

The amount of K-feldspar is less in the deeper than the shallow samples and is a likely source of K for the formation of illite. K-Ar studies of size fractions indicate this is probably true (Weaver and Wampler, 1970).

Van Olphen (1966) heated K-saturated montmorillonite to ~ 400° C and was able to obtain a maximum contraction of only 50 percent of the layers; K-saturated montmorillonite heated at 400° C under 500 atm collapsed to 10.3 Å (Khitarov and Pugin, 1966), which indicates approximately 25 percent of the layers are not contracted to 10 Å. Winkler (1967, p. 174) reports that Althaus obtained an ordered paragonite-montmorillonite mixed-layer structure, from Na montmorillonite, at temperatures of 370° C and 2000 bars water pressure. It is apparent from these experiments as well as from theoretical considerations that "typical" montmorillonite cannot be converted completely to illite at reasonable temperatures and pressures without the addition of other elements, notably Al. Thus after the initially high charged layers have contracted, during burial, some chemical change must occur to allow for additional permanent contraction. LiCl treatment (Green-Kelly, 1955) of shallow Springer Shale and Chevron Oil Company samples and samples from the bottom of the Gulf of Mexico indicate the swelling clays are largely beidellitic. Thus, they have a relatively high tetrahedral charge, but the total charge is low.

At shallow depth (less than ~ 10,000 ft), in areas where the geothermal gradient is relatively low, it seems unlikely that much structural rearrangement occurs. The interlayer hydroxides probably become more stable, and the K becomes concentrated in the highest charged layers. Further collapse apparently requires some changes which affect the basic lattice. The formation of additional chloritic material requires temperatures high enough to cause the dissolution of kaolinite.

A series of Chevron samples was treated with tetraphenolboron for varying periods of time to determine the ease and rate of extraction of K. These experiments (to be reported in another paper) show that, with increasing depth of the sample, the K is more difficult to extract. This indicates that an increase in lattice charge probably occurs with increasing depth.

A process that occurs early, at least in the Chevron samples, is the reduction of Fe^{3+}. At 4233 ft the amount of Fe^{2+} is already abnormally high and increases with increasing depth. If a fair portion of the reduced Fe is in the octahedral layers of the montmorillonite, then the charge of some of these layers could be increased enough to allow the fixation of additional K. Dioctahedral octahedral layers can contain a maximum of 0.2 parts Fe^{2+} per two positions (Weaver, 1968). Some Fe^{2+} is present in siderite (and chlorite), but Mossbauer curves indicate this siderite content is relatively small.

Diagenesis that involves changing the composition of the clay minerals requires temperatures higher than those encountered at the bottom of the Chevron well (119° C). Higher temperatures are encountered in the deep wells penetrating the Springer Shale (similar to 140° C at 20,000 ft and 170° C at 24,000 ft—Levorsen, 1967, p. 420).

At these temperatures and pressures (and presumably basic conditions), kaolinite is unstable and is destroyed. Some of it may convert to dioctahedral chlorite, but it seems more likely that the Al is deposited as gibbsite sheets in some of the expanded layers of the mixed-layer phase. When a sufficiently large number of layers have acquired gibbsite sheets, they diffract as a discrete chlorite phase. A weak peak at 4.03 Å on many of the X-ray patterns suggests that some of the silica forms cristobalite.

At some stage it is necessary to increase the amount of tetrahedral Al in the 2:1 montmorillonitic layers in order to form illitic layers.

In addition to the kaolinite-Al-forming gibbsite sheets, Al^{3+} ions + H_2O probably would be present between some of the layers. When the temperature becomes high enough, the interlayer Al^{3+} could migrate conceivably into the tetrahedral layer in a manner analogous to Li. Tettenhorst (1962) has shown that when montmorillonite is heated at 300° C, small ions such as Li^+, Mg^{2+}, and Ni^{2+} will migrate from the interlayer area to a position in the hexagonal holes of the tetrahedral layer of montmorillonite. The presence of Al^{3+} in the interlayer area should facilitate the development of a high-Al tetrahedral layer. It is possible that the presence of Al^{3+} in the hexagonal cavities can increase the rotation of the silica tetrahedron to the extent that the Al^{3+} could capture some of the basal oxygens and form Al tetrahedrons (*see* Appendix by C. O. Pollard, Jr.). Some silica would be lost from the layer, creating the necessary charge deficiency. Radoslovich and Norrish (1962) have noted that a certain amount of misfit between the size of the tetrahedral layer and octahedral layer is necessary to allow tetrahedral rotation and the locking in of the K ion. Under low-temperature conditions it would seem that changes in the flexible tetrahedral layer would be easier than in the relatively rigid (and less accessible) octahedral layer. Further, the octahedral layer of montmorillonite is similar to that of illite.

Another possibility is that Al and Si tetrahedra form at the edges of the montmorillonitic flakes. This would result in the formation of a flake with a montmorillonitic core and an illitic rim. Both mechanisms may be operative. The latter is less likely.

The development of an illite-rich clay suite from the starting materials present in the Pliocene-Miocene of the Gulf Coast and in the Springer Shale of Oklahoma could be hindered by a lack of Al, K, and temperature. The availability of Al is controlled by kaolinite and feldspar; K, largely by K-feldspar; and heat, by depth, geographic location, water content, and rate of deposition.

The Pliocene-Miocene mud and Springer Shale samples are apparently deficient in K and temperature, and perhaps Al. The Al_2O_3 from 10 percent kaolinite (of the clay suite) would amount to approximately 4 percent, and from 15 percent kaolinite Al_2O_3 would amount to 6 percent. If the initial clay suite contained 60 percent expanded layers with an Al_2O_3 content of 25 percent (relatively high) then 6 percent Al_2O_3 would be needed to convert these layers to muscovite-like layers ($\sim$ 35 percent Al_2O_3). Even though some of the Al is used to form chlorite layers, it would seem that a K deficiency is more a factor than an Al deficiency. Temperature is a factor, but composition will influence the temperature at which changes occur. If the composition is adequate (sufficient Al and K), the transformation of montmorillonite to mica-like layers can take place at temperatures less than 200° C. If the sample is low in Al or K, or both, complete conversion will not occur until temperatures of 400° C and larger are reached (Winkler, 1967, p. 175). At this stage, structural reorganization and the formation of new phases occur.

Low-Grade Metamorphism

The next stage in the diagenesis of clay minerals can be seen best in an area of regional dynamothermal metamorphism. The boundary of the disappearance of kaolinite in the metamorphosed Ouachita sediments of Texas closely coincided with the isothermal gradient contours. The thermal gradient in the area indicates temperatures of 150° C exist at approximately 10,000 ft. Kaolinite is destroyed at a lower grade of metamorphism than is the mixed-layer 2:1 material (Weaver, 1960). The 10 Å illite-mica peak becomes progressively sharper as the degree of metamorphism and temperature increases. These lateral clay mineral changes are more likely due to an increase in temperature rather than pressure. The lateral changes are presumably equivalent to the vertical changes occurring in the Tertiary and Springer sediments.

The 2:1 clay mineral consists of illite and a mixed-layer illite-chlorite-montmorillonite (all presumably dioctahedral). During the earlier stages (incipient metamorphism), the expanded montmorillonite layers are lost. Al freed from the kaolinite is probably deposited in the interlayer position, increasing the chloritic component of the mixed-layer illite-chlorite.

The mixed-layer illite-chlorite is indicated by the asymmetry of the 00ℓ illite peaks—the low-angle side of the 10 Å peak, the high-angle side of the 5 Å peak. As the relative intensities of the 5 Å and 4.75 Å peaks are more nearly equal than the 10 Å and 14 Å peaks, the presence of the chlorite is best seen in the former. The chlorite is presumably dioctahedral in both the pyrophyllite and gibbsite layers.

The 14.2 Å peak and the 4.75 Å in many of these samples show an asymmetry toward the 10 Å and 5 Å illite peaks, respectively, suggesting that some illite is interlayered with the chlorite component. As metamorphism and temperature increase, the mixed-layer phases are exsolved, and discrete sharp illite (mica) and chlorite peaks are obtained.

Micas in the lower grades of metamorphic rocks have compositions similar to those of the higher charged illites (Bates, 1947; Lambert, 1959; Ernest, 1963; Van der Plas, 1969) and are called phengites. If it is true that much of the compositional difference between illite and muscovite is due to the presence of interlayered dioctahedral chlorite, then it is likely that the chlorite is still present in these low-grade metamorphic micas (perhaps at this stage, the hydroxyl water is removed from the gibbsite layer, and an Al mica is formed). The high-grade metamorphics contain micas more similar to muscovite indicating exsolution is nearly complete, and pure muscovite and chlorite layers are created. Hydrothermal experiments by Winkler (1964) and Velde (1964), using natural illites, showed the following transitions; illite ⟶ chlorite + muscovite ⟶ biotite + muscovite. The illitic starting materials presumably contained chloritic layers.

Previous Studies

Warshaw's (1959) hydrothermal studies showed that it was virtually impossible to produce a single 10.0 Å mica phase, below 400° to 500° C, when the composition of the starting material varied much from the ideal muscovite. Velde (1964) substantiated this and reported that a mixed-layer illite-chlorite was formed at low temperatures (at even lower temperatures, that is, less than 300° C, expanded clays were formed). He suggested that most illites consist of interlayers of chlorite and a mica layer similar to muscovite in composition. Rex (1964) presented data suggesting some chlorite-like layers occurred in authigenic illites (10.4 Å) formed in sandstones.

Weaver (1965) reported on the asymmetry of the 00ℓ peaks of Paleozoic and Precambrian illites and concluded that some 14 Å to 15 Å layers must be interlayered with the 10 Å illite layers. A re-examination of the X-ray patterns shows that there is a good linear relation between the width at half-height of the 10 Å and 5 Å peaks and the K_2O content. The 10 Å peak doubles in width as the K_2O content decreases from 10.5 to 8 percent. This increase in width is presumably due to the presence of chlorite interlayers. The 10 Å peak widens on the low-angle side and the 5 Å peak on the high-angle side. This is further indication that 14 Å layers are present. Raman and Jackson (1966) fairly convincingly proved the broadening was largely due to chlorite layers (varying amounts of vermiculite and montmorillonite layers were also present). They concluded that the illites they examined contained only from 58 to 65 percent 10 Å layers and from 20 to 29 percent chlorite layers.

Dioctahedral chlorites are common (for review *see* Rich, 1968), and are relatively stable under a variety of conditions. Thus it might be expected that mixed-layer illite-chlorite-montmorillonite and illite-chlorite would be relatively common. Because of the X-ray characteristics of such mixtures, they are not recognized easily.

Interstitial Ions

CATIONS

Na is the major cation in the interstitial water, comprising over 90 percent of the cation suite for most samples. K is second in abundance, followed by Mg and Ca. The vertical distribution of the K/Na, Mg/Na, and Ca/Na ratios are shown in Figure 6.

The Ca/Na ratio values are relatively uniform to a depth of approximately 12,000 ft where they decrease abruptly and remain relatively constant to total depth. The K/Na values are similar to seawater to a depth of approximately

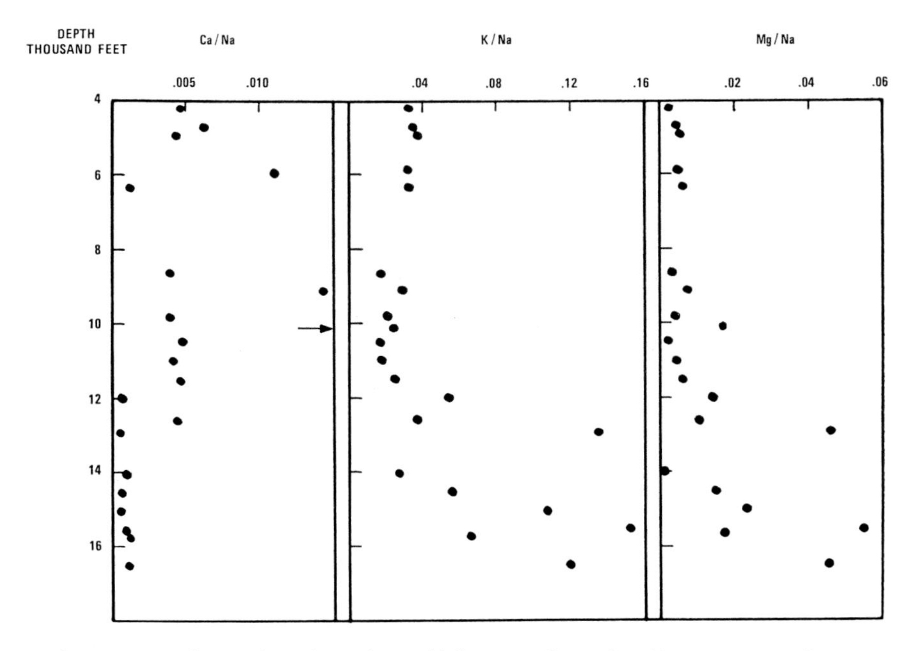

Figure 6. Cation ratios of the interstitial waters from the Chevron cores. Seawater ratios are Ca/Na = .038; K/Na = .036; Mg/Na = .12.

6000 ft. From 8600 ft to 12,000 ft the values decrease slightly, then increase. Values below 12,000 ft are two to four times larger than those of seawater. The Mg/Na values are relatively uniform to a depth of 12,000 ft. Deeper samples have larger ratios which, in general, increase with increasing depth. Mg/K and Ca/K ratios are much less than for seawater. The former remain relatively constant with depth; the latter decrease with depth. The Mg/Ca ratio is less than seawater to a depth of 12,600 ft; deeper samples have values much larger than seawater.

An attempt was made to obtain some absolute values by calculating the concentration of interstitial cations per gram of water lost from the samples on drying at room temperature (Fig. 7). Similar distribution patterns were obtained by calculating grams of cation per grain of dry sample. The vertical variation for Ca, K, and Mg is very similar to that for Ca/Na, K/Na, and Mg/Na suggesting that the variation is not solely due to the variation in Na. The Na values systematically increase to a depth of approximately 11,500 ft and then decrease.

The total cation population systematically increases to approximately 10,000 ft, then abruptly decreases by about 20 percent and in the deeper samples remains relatively constant or may increase slightly. The cation concentration ranges between approximately 15,000 to 25,000 ppm or an average of approximately twice that of seawater. If these values have any validity, they indicate that Na and K are consistently more concentrated than in seawater, and the Ca and Mg values are less than for seawater.

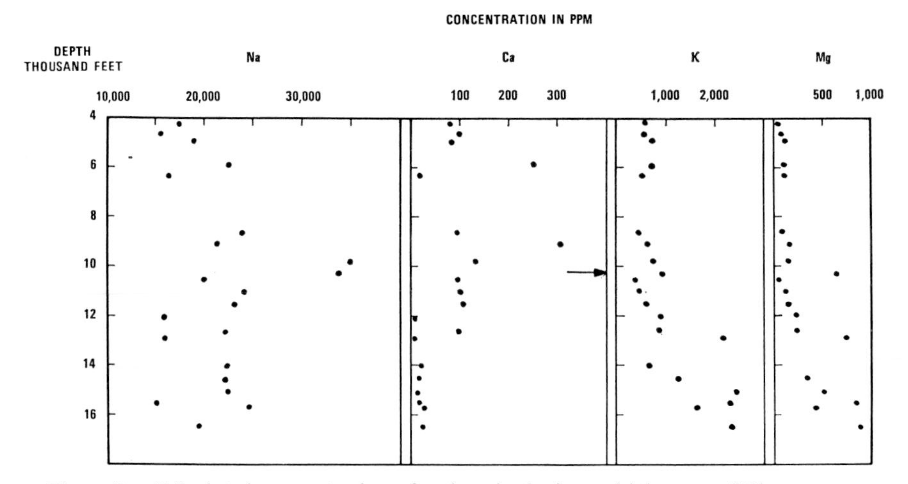

Figure 7. Calculated concentration of cations in the interstitial water of Chevron cores. Based on water loss at room temperature. Values for seawater are: Na = 10,560; Ca = 400; K = 380; Mg = 1272.

Figure 8 is a plot of the percent water loss of samples (at room temperature) *versus* depth. There is a systematic decrease to 10,300 ft, then an abrupt increase, followed by a systematic decrease. The abrupt increase in pore water between 10,300 and 10,530 ft coincides approximately with the occurrence of abnormally high fluid pressures encountered at 10,230 ft. The low values for the 9800 and 10,300 ft samples reflect the presence of a permeability barrier or "cap" in this interval.

Kerr and Barrington (1961) and Myers and Van Siclen (1964) found that in samples from a number of Gulf Coast wells the bulk density systematically

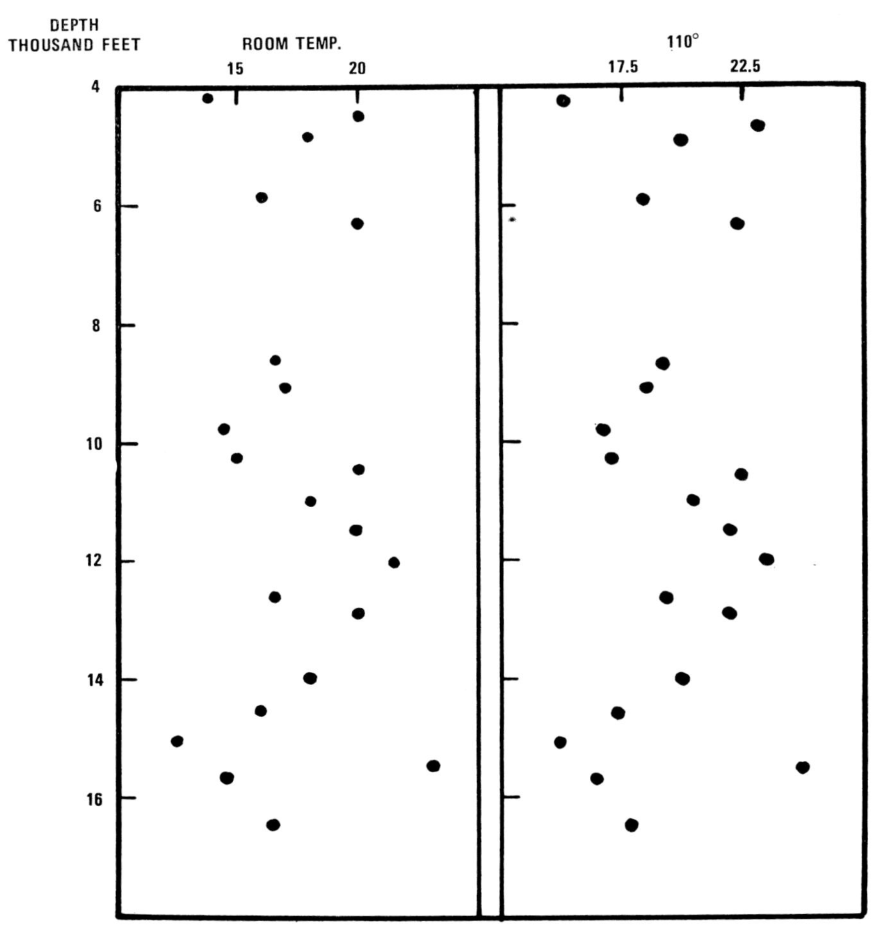

Figure 8. Percent of water lost from Chevron cores at room temperature (~25° C) and at 110° C. Calculated as percent of moist sample.

increased to depths ranging from 8000 to 14,000 ft. Within this depth interval there is an abrupt decrease in bulk density followed by a systematic increase. This reversal takes place over an interval of less than 500 ft. In some of the wells, high interstitial fluid pressures occur in the sands associated with the low-density shales.

The reversal in density is presumably due to an increase in the amount of interstitial water, as was noted in the well investigated for the present study, and will be discussed in more detail in the section on compaction.

A comparison of Figures 7 and 8 indicate that in the shallower samples the concentration of Na (based on ppm Na per gram of water lost by drying at room temperature) in the interstitial waters increase as the amount of water decreases to a depth of 10,300 ft. There appears to be a major increase in the concentration of Na in the two deepest samples with the minimum water content. There is a decrease in the Na concentration at 10,530 ft coinciding with the increase in the amount of interstitial water. From 10,530 ft and deeper, the Na concentration remains relatively constant, but may decrease as the amount of interstitial water decreases. Overton and Zanier (1971) found a similar vertical distribution for Na in Gulf Coast overpressured formations.

The total amount of Na (calculated as gm Na per gram of sample dried at 110° C) has a similar distribution (Fig. 9). Below 10,530 ft the Na content varies directly as the amount of interstitial water varies, and above 10,530 ft there is an inverse relation. A plot of ratio grams of interstitial Na/percent Al_2O_3 shows the same relation.

Values of *p*H were determined on the water used to recover the interstitial ions (100 ml of distilled water to 10 gm clay). Values ranged from 8.4 to 8.9; the large majority were very close to 8.7 (Table 3). There is little variation with depth, though the samples shallower than 6000 ft may have slightly lower values.

The *p*H values are of questionable significance, but may be reasonable. Both values of *p*H that are higher (Rittenberg and others, 1955) and lower (Siever and others, 1965) than seawater have been reported for shallow marine muds. The *p*H of the drilling fluid ranged from 10 to 11 and suggests that contamination from the drilling fluid was probably minor.

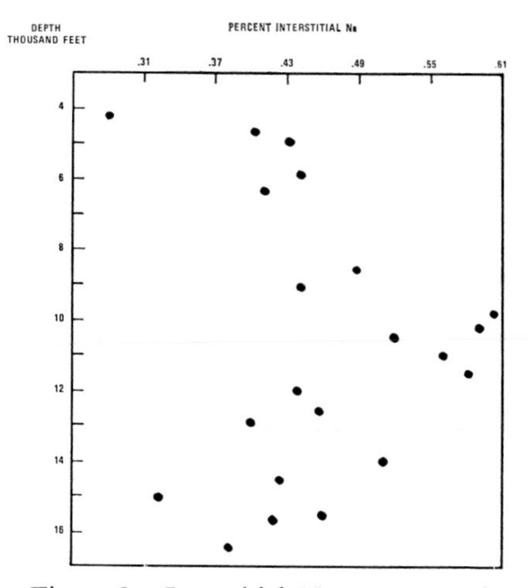

Figure 9. Interstitial Na as percent by weight of bulk Chevron sample dried at 100° C.

TABLE 3. WATER COLOR AND *p*H OF INTERSTITIAL WATERS

Depth	*p*H	Water Color (Clay Settled)
4710	8.40	colorless
4950	8.64	light brown
5900	8.45	colorless
8650	8.55	faint yellow
9100	8.665	faint brown
9800	8.88	light brown
10530	8.61	light brown
11020	8.505	medium brown
11520	8.910	medium brown
12610	8.71	medium brown
12890	8.535	dark brown
14010	8.745	dark brown
15044	8.705	medium brown
15518	8.590	very dark brown
15650	8.64	medium brown

ANIONS

HCO_3 is the dominant anion in the interstitial waters; SO_4 is next in abundance, and is followed by Cl. As expected, the distribution of the anions is similar to that of the total cations—systematically increasing to approximately 11,000 ft and then systematically decreasing (Fig. 10).

The equivalent parts per million (epm) values for the cation and anion suite of each sample are quite similar (Table 4) considering the possible errors. For all samples, but one, that do not balance, the anion values are lower than the cation values, and may be due to the presence of CO_3^{-2} or functional organic groups.

The amount (weight per gram sample dried at room temperature) of Cl decreases fairly regularly from 4000 to 16,450 ft, although there is a slight reversal of trend between 9000 and 11,000 ft. SO_4 systematically increases to a maximum at 110,000 ft, then systematically decreases. The HCO_3 values decrease sharply from 4000 to approximately 7000 ft. A trend with less steep slope can be extended to 10,000 ft. Values are scattered for the deeper samples, but tend to decrease with increasing depth. The sample from 10,300 ft has abnormally high values for Cl and SO_4 and a low value for HCO_3.

The cation and anion values for the 10,300-ft sample were so anomalous that, initially, some of the data were not recorded. The values reported are based on extrapolation of data obtained from analyses of a series of dried cores.

The concentration values for anions (calculated as ppm/gm water lost at room temperature) have similar distribution patterns (Fig. 11).

Thus, Cl is systematically removed from the interstitial water; HCO_3 and

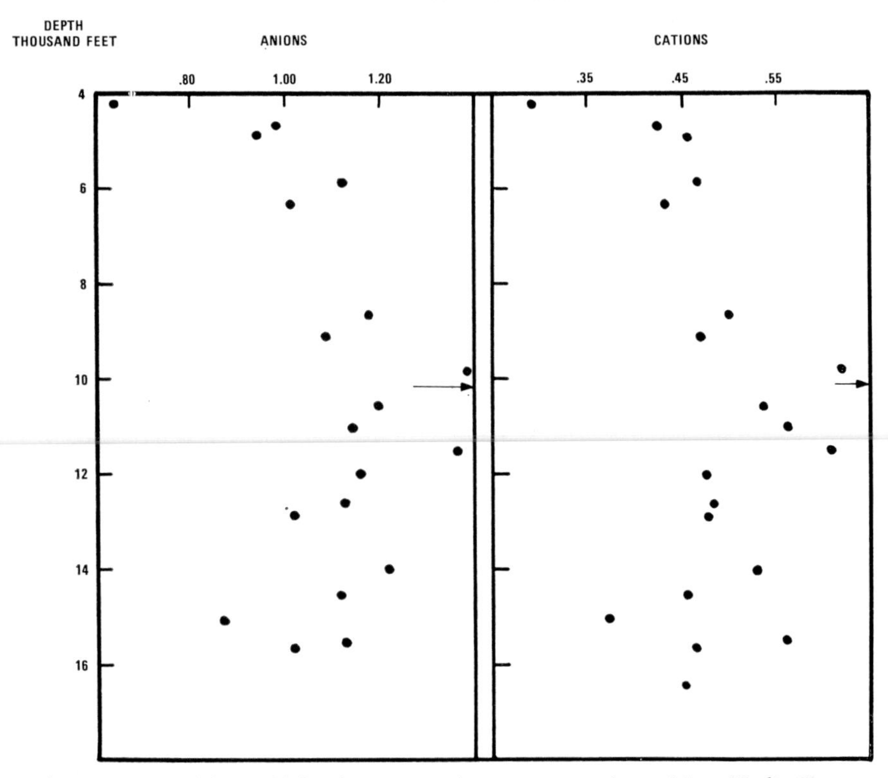

Figure 10. Total interstitial anions and cations as percent by weight of bulk Chevron sample dried at 110° C.

SO_4 values systematically increase to approximately 10,000 ft, and then remain relatively uniform.

Tageyeva (1965) studied the interstitial waters of Holocene muds from the Arctic Ocean and found that both SO_4 and HCO_3 were usually more abundant than in seawater, whereas Cl was less abundant. HCO_3 concentrations in the interstitial waters are as much as 20 times higher than seawater, and SO_4 values nearly three times higher. Cl concentrations are as low as half that of seawater.

Analyses of sandstone formation waters from the Mississippi Delta area (Jones, 1968, p. 12) indicate that bicarbonate is from 2 to 20 times as abundant as in the Gulf of Mexico water. Sulfate is consistently less by a factor of 4 to 230.

Assuming the initially trapped interstitial water had the Cl content of seawater (18,980 ppm), then by a depth of burial of 10,000 ft the Cl concentration

TABLE 4. EQUIVALENT PARTS PER MILLION OF INTERSTITIAL WATERS

Cation Suite	Depth in Feet									
	4233	4710	4950	5900	6350	8650	9100	9800	10300	10530
Na	10.3	13.6	15.2	15.8	14.1	17.1	15.8	21.8	. . .	17.6
Ca	0.05	0.1	0.08	0.2	0.02	0.08	0.3	0.1	. . .	0.1
K	0.2	0.3	0.3	0.3	0.3	0.2	0.3	0.2	0.3	0.2
Mg	0.5	0.1	0.2	0.1	0.2	0.1	0.2	0.2	0.8	0.07
Total	10.60	14.1	15.78	16.4	14.62	17.48	16.6	22.3		17.97
Anion Suite										
Cl	3.8	3.9	4.4	3.1	3.0	2.6	1.2	3.0		2.3
SO_4	1.6	5.7	2.7	2.1	2.9	4.8	6.1	5.3		5.8
HCO_3	(5.2)*	5.2	7.6	11.2	(8.7)	10.0	8.6	12.7		9.0
Total	10.6	14.8	14.7	16.4	14.6	17.4	15.9	21.0		17.1
Cation Suite	Depth in Feet									
	11020	11520	12020	12610	12890	14010	14538½	15044	15818	15650
Na	18.9	19.9	14.8	16.1	13.8	17.8	15.3	12.1	15.2	15.4
Ca	0.09	0.1	0.01	0.08	0.01	0.02	0.01	0.01	0.02	0.02
K	0.2	0.3	0.5	0.36	1.1	0.3	0.5	0.8	1.4	0.6
Mg	0.2	0.2	0.4	0.3	1.2	0.1	0.4	0.5	1.6	0.5
Total	19.39	20.5	15.71	16.84	16.11	18.22	16.21	13.41	18.22	16.52
Anion Suite										
Cl	3.1	2.7	1.5	1.7	1.7	1.4	1.3	0.9	1.3	0.9
SO_4	6.3	4.7	3.9	5.0	5.0	5.1	4.2	3.4	5.2	6.1
HCO_3	7.8	11.9	(10.3)	9.6	7.8	10.7	(10.7)	7.8	8.8	8.2
Total	17.2	19.3	15.7	16.3	14.5	17.2	16.2	12.1	15.3	15.2

* Calculated

had decreased by a factor of 4 to 5 (4000 ppm); concurrently the SO_4 concentration increased by a factor of 6 to 7 (2649 ppm to 17,000 ppm). White (1965) and others have noted that the SO_4/Cl values (0 to 0.05) of connate waters in most permeable sedimentary rocks are usually considerably less than for seawater (0.14). The ratio values in the mud samples range from 0.6 to 10. If all the SO_4 is concentrated by selective filtration, then the interstitial water content has decreased by a factor of 7. If the present water content is 15 percent by weight of wet samples, then the original water content would have been approximately 100 percent. This is a reasonable value. Evidence has been presented (review *by* White, 1957) to show that the sulfate in interstitial waters in muds is generally reduced to sulfide by the action of sulfur-reducing bacteria. The bacteria reduce sulfate by utilizing energy available from oxidation of organic material to CO_2 and other substances. This does not appear to be the case for the present samples. Either sulfate is produced, or it is con-

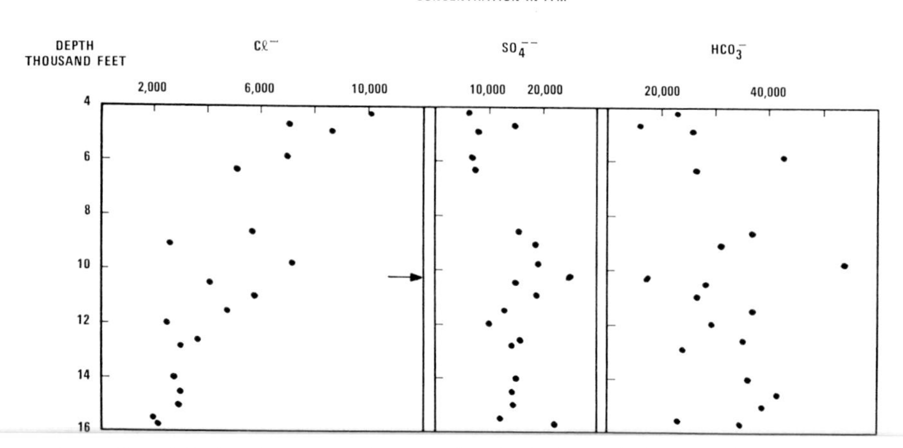

Figure 11. Calculated concentration of anions in the interstitial water of Chevron cores. Based on water lost at room temperature. Values for seawater are: Cl = 18,980; SO_4 = 2649, HCO_3 = 140.

centrated by a filtration effect. White (1965, p. 351) has suggested that because of its size SO_4 is much less mobile than Cl. Yaalon (1965) reported that, in soils, Cl migrates more readily than SO_4.

The growth of bacteria may be retarded by various cations (Porter, 1946) and by increased salinity. Kuznetsova (1960) thought the lack of sulfate reduction in some ground waters was due to the inhibiting effect of bivalent cations. The extremely rapid rate of deposition in the deltaic area may be the major factor in limiting the amount of sulfate reduced. It is not axiomatic that sulfate be reduced to sulfide in marine muds.

As White (1965) and others have noted, the SO_4/Cl values of most porous reservoir waters (sandstones and limestones) are less than that of seawater. The present data suggests that some of this may be due to selective filtration rather than sulfate reduction.

Ocean waters contain an average of 140 ppm HCO_3. The high concentration of HCO_3 in the Chevron samples (20,000 ppm at 4000 ft to 40,000 at 10,000 ft) cannot be entirely due to selective filtration. Bicarbonate and free CO_2 are products of the decomposition of organic matter. It is believed (White, 1957) that sulfate-reducing bacteria, which obtain this energy by the oxidation of organic matter, play a major role in the production of CO_2. The high content of SO_4 in the present samples suggests that a different mechanism must be operative. Ferric iron may be a source of oxygen. The ferrous iron content of the Chevron samples averages over 2 percent (Table 7). Iron-reducing bacteria could obtain from the ferric iron the oxygen necessary to oxidize the organic matter.

Surface marine sediments from this area of the Gulf of Mexico contain from 6 to 16 percent calcite (Shaw and Weaver, 1965), yet only minor amounts of calcite were found in three of the Chevron samples. A small amount of the calcite may be dolomitized, but much of it must go into solution and be removed from the mud system. This must be the source of much of the Ca concentrated in the connate waters in sandstones. The calcite also could be the source of much of HCO_3 in the interstitial waters of the mud samples.

CO_2 produced by the breakdown of the organic material would cause the partial solution of the calcite. Some of the Ca would be adsorbed on the clays and some would be squeezed from the interstitial water of the muds. The carbonate ion will react to form bicarbonate. All three mechanisms would favor the continued dissolution of the calcite. Humic and fulvic acids, which are present in the interstitial water, can also dissolve calcite.

Interstitial waters of sandstones associated with marine muds and shales generally have a lower Na/Cl, K/Cl, Mg/Cl, HCO_3/Cl and SO_4/Cl ratios and a higher Ca/Cl ratio than seawater. The interstitial waters in the muds, examined in this study, have ion/Cl ratios which are essentially the opposite, with respect to seawater, of those found in sand waters. All the ion/Cl ratios except Ca/Cl and Mg/Cl are significantly larger than seawater. Na/Cl ratios attain values on the order of 20 times greater than seawater; K/Cl, 50 times; SO_4/Cl, 70 times; and HCO_3/Cl, 2300 times. The low absolute value for Cl indicates that the high ratios are partially due to the high mobility of Cl, but only in part do these values indicate the ease of migration of these various ions.

Exchange Cations

Ba was used to replace the exchangeable cations. The amount of exchangeable Na, K, Ca, and Mg was determined. The total cation exchange capacity was not measured, but should not differ much from the values obtained by adding the four cations.

Total exchangeable cations (Na + K + Ca + Mg) tend to decrease with depth (Fig. 12). The shallower samples (above 8000 ft) have values that average slightly larger than 30 meq/100 gm; the deepest samples average close to 20 meq/100 gm. Abnormally high values occur at the top of the high pressure interval (10,300 and 10,530 ft), and are probably due to the presence of the small amounts of calcite (X-ray) in these samples. A comparison of the C.E.C. and Al_2O_3 values indicates that the variation in C.E.C. is primarily due to variations in the relative proportion of contracted and expanded 2:1 layers.

The C.E.C./Al_2O_3 (Fig. 13) values decrease continuously down to approximately 10,000 ft; deeper samples have a uniform ratio. Presumably, these data reflect the change in the relative proportion of expanded layers with depth

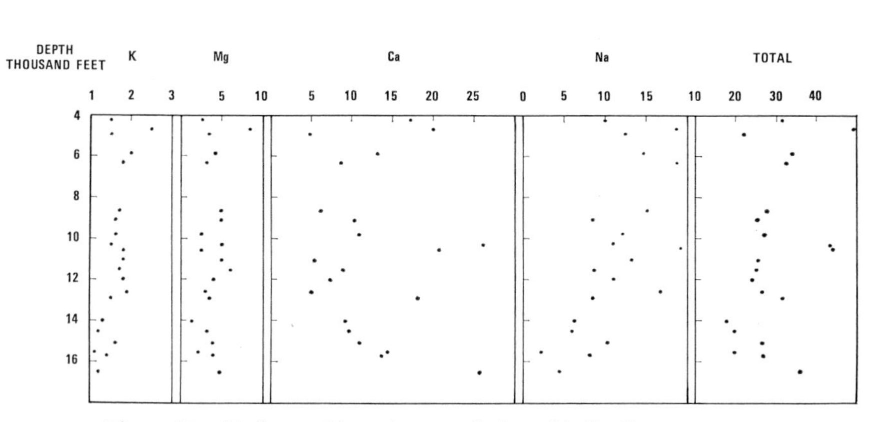

Figure 12. Exchangeable cation population of bulk Chevron cores.

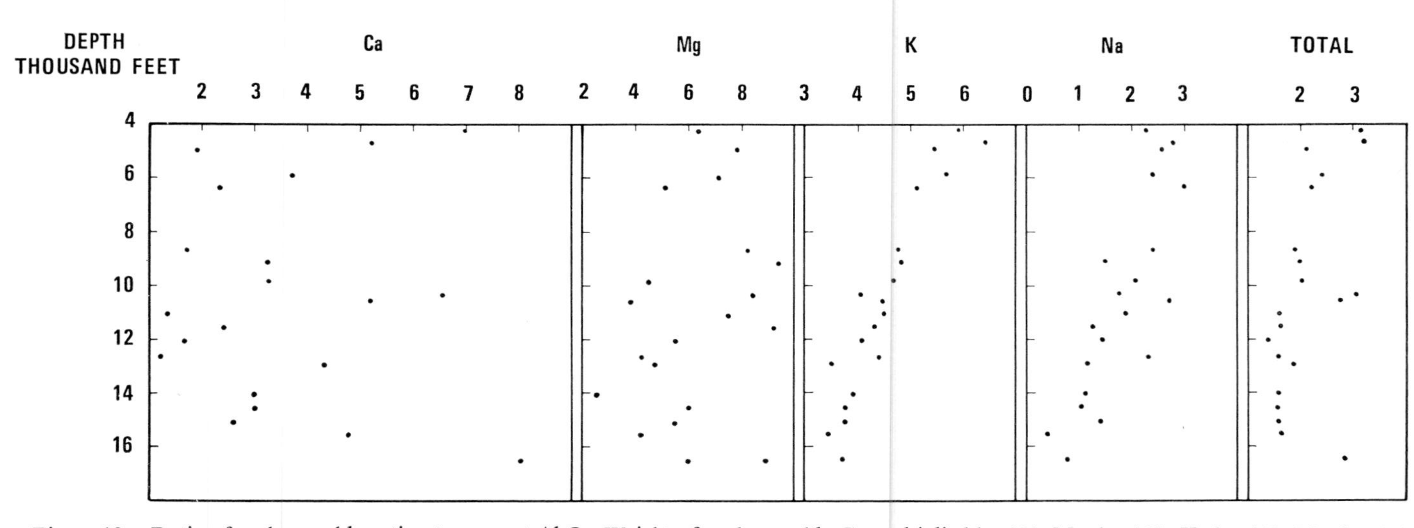

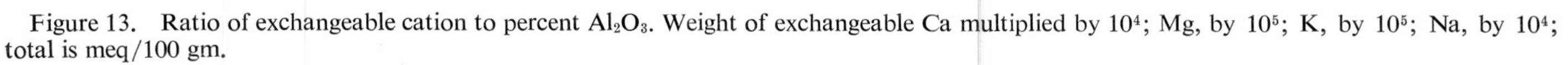

Figure 13. Ratio of exchangeable cation to percent Al_2O_3. Weight of exchangeable Ca multiplied by 10^4; Mg, by 10^5; K, by 10^5; Na, by 10^4; total is meq/100 gm.

and indicate that the rate of loss of interlayer water is greatest above the permeability barrier.

Plots of the ratio of exchangeable cation (weight per gram) to Al_2O_3 *versus* depth (Fig. 13) show that the exchangeable K/Al_2O_3 and exchangeable Na/Al_2O_3 ratios systematically decrease with depth. The exchangeable Mg/Al_2O_3 values are more scattered, but show the same relation. The exchangeable Ca/Al_2O_3 values are widely scattered, but there is some suggestion that the ratio increases with depth. It seems likely that some of the "exchangeable" Ca came from the carbonate minerals.

A plot of the exchange cation/interstitial cation ratios (weight per gram dried at 110° C) *versus* depth (Fig. 14) shows that the ratio values for Na, K, and Mg decrease with depth, with most of the decrease occurring below approximately 12,000 ft. Ca has an opposite trend with most of the increase occurring below approximately 12,000 ft. Na is the only cation more abundant in the interstitial water than in the exchange position. Exchangeable K is 1 to 8 times as abundant as interstitial. As exchangeable Mg and Ca are much more abundant than interstitial Mg and Ca, a preference of the clay minerals for the divalent cations, particularly Ca, is suggested.

Published data indicates the exchangeable cation suite on clays deposited in seawater varies considerably. However, most data indicate that Mg is the major exchange cation on montmorillonitic clays; K is the least abundant;

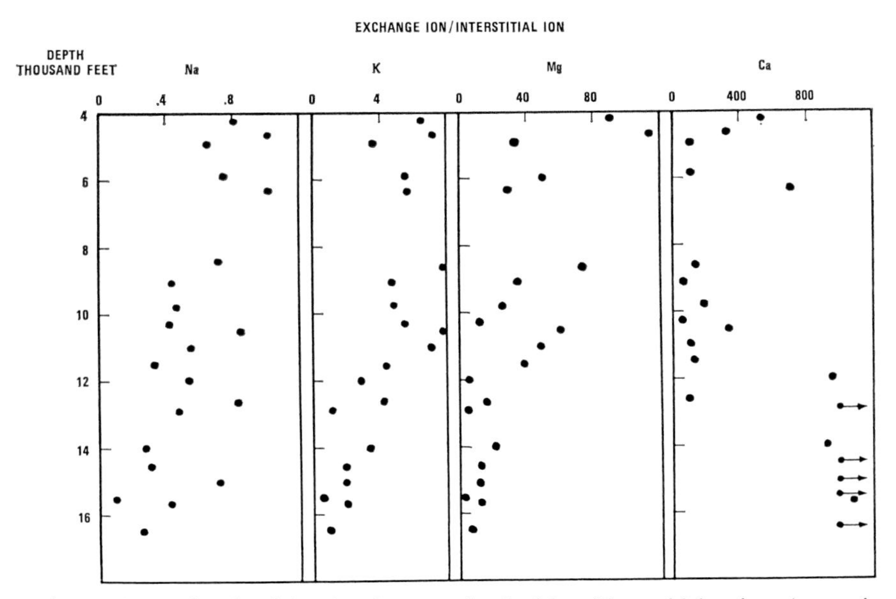

Figure 14. Ratio of weight of exchange cation/weight of interstitial cation. Arrows in Ca graph indicate values ranging from 1090 to 7300.

Na and Ca are present in variable amounts, generally less than Mg and more than K.

Thus, it would appear that between the surface and 4000 ft, exchangeable Mg is lost, apparently replaced by Ca and Na. The cation suite is relatively constant between 4000 ft (and probably shallower) and approximately 12,000 ft. Below this depth, exchangable Ca increases at the expense of Na.

As the Mg content of the interstitial waters in the Chevron muds and in most marine sands is less than in seawater it would appear that the original interstitial Mg plus the exchangeable Mg is incorporated, in a nonexchangeable form, in other minerals.

If it is assumed that half the original exchange cations were Mg and that the original surface mud was composed of equal parts by weight of sediment and seawater, then approximately 0.6 percent Mg (1.02 percent MgO) was present (based on weight of sample dried at 110° C). The samples at depth contain approximately 0.1 percent Mg in the exchange position and in the interstitial water. The bulk samples contain on an average 1.66 percent MgO. It seems unlikely that much of the nonfixed Mg has remained in the mud. Studies of Tertiary marine sands from the Gulf Coast by the senior authors have shown that poorly crystallized chlorite and mixed-layer chlorite-montmorillonite are commonly a major component of these sands. Thus, it is suggested that some of the loosely held Mg is removed from the muds and migrates to the sand where it reacts with montmorillonite to form a chlorite-like clay.

The X-ray and chemical data indicate that minor amounts (probably less than 5 percent in most samples) of dolomite are present in all samples. It is possible that a portion of the available Mg was used to form the dolomite. After removal of the exchange cations, the noncalcite samples contain from 0.4 to 1.3 percent CaO. Most of the Ca must be in the dolomite. Therefore, the amount of dolomite is sufficient to account for much of the lost exchangeable Mg.

With depth, 4000 to 16,000 ft, the percentage of Ca in the exchange cation suite increases, and the percent Na decreases; however, Ca as weight of total sample remains relatively constant, whereas the amount of Na decreases (from 0.3 to 0.4 percent to 0.1 to 0.2 percent). Thus, as the expanded layers, particularly below 11,000 ft, contract the interlayer, Na cations are released preferentially. The composition of the cation suite in the deepest samples essentially is identical to the suite commonly found on clay in fresh-water rivers.

The amount of exchangeable K is quite low (Fig. 12). It comprises from 5 to 7 percent of the exchange cation suite (1.5 to 2.0 meq/100 gm) to 13,000 ft, and then decreases slightly. The concentration of K in the interstitial waters also remains relatively uniform, and greater than seawater, until approximately 13,000 ft, then increases. The decrease in exchangeable K (in weight per gram 110° C) is roughly similar to the amount of increase in interstitial K.

It might be expected that K would be extracted from the interstitial waters to facilitate the contraction of the expanded clay layers. This does not seem to be the case. From the surface (seawater) to 10,000 to 11,000 ft, the concentration of both Na and K approximately doubles. This could be interpreted to mean that Na and K left the interstitial mud waters at the same rate; however, as the K/Na ratio of sand waters is less than seawater, the implication is that K is taken up by the clays. The amount of K adsorbed or fixed by the clays would be small. In the deeper samples it would appear that K is definitely released from the clay minerals or the K-feldspars, or both.

The concentration of K in the interstitial waters down to 14,000 ft is similar to that found by Siever and others (1965) for shallow marine muds. Thus, the K concentration (slightly less than twice that of seawater) presumably remains relatively constant from the surface to 14,000 ft and then increases. This suggests that equilibrium conditions exist between the interstitial waters and the illite and K-feldspar in the muds. Below 14,000 ft, the increased K and K:Na ratios may reflect the increase in temperature. White (1965, p. 358–360), has noted that the K:Na ratio of waters in equilibrium with sediments containing K-micas and K-feldspars should increase with increasing temperature.

If it is assumed that the surface muds contain equal parts by weight of water and solid, the amount of K would be approximately 0.0006 gm K/gm solid (using data *in* Siever and others, 1965). From the surface to 10,000 ft (and 16,000 ft), the water content is decreased by a factor of approximately 7, with the K concentration remaining constant (from 4000 to 14,000 ft the amount of interstitial K is approximately 0.00015 gm/gm solid). Thus, approximately 0.00045 gm K/gm solid is lost from the interstitial water. This is equivalent to 0.05 percent K_2O. Some of this K is flushed from the muds, but even if all of it were fixed between montmorillonite layers, it would allow the formation of only 0.5 percent of illite layers.

After the Chevron core samples had dried at room temperature, the base exchange cations were determined in the same manner as before. The total cation exchange capacity (K + Na + Ca + Mg) is less for all samples, except one. The decrease ranges from 1 to 29 meq/100 gm and averages 11 or approximately one-third the exchange capacity of the nondried samples.

Even though the total values are low the percentage composition of the exchange cation suite of the dried samples is nearly identical to that of the wet samples. The reason for this is not known, but may indicate incomplete dispersion of the dried samples rather than any cation fixation.

MECHANISM

In the shallow samples, the exchange suite is Na > Ca > Mg > K; in the deepest samples, the exchange suite is Ca > Mg and Na > K; by extrapolation, the suite at ~ 17,000 ft would be Ca > Mg > K > Na. Thus, with depth, the increase in the proportion of exchange cations is Ca > Mg > K > Na. The ratio of weight exchange cations to weight interstitial water cations shows the following series in order of decreasing difference: Ca > Mg > K > Na. Na is the only ion with a ratio less than one. This latter series is identical to the one Ross (1943) gave for the order of decreasing replaceabliity in montmorillonites. Marshall (1954) has shown that the order of bonding energies ($\triangle F$) for beidellite and illite are Ca > Mg > K > Na. The order of Ca and Mg is reversed for Wyoming bentonite. Thus, with increasing depth of burial the exchangeable cations are systematically removed from the exchange positions in the general order of relative ease of replacement (and suggests also that the exchange data is real).

The fact that the amount of dominance of exchange cations over interstitial cation decreased in the same series sequence as the ease of replaceability series suggests that the exchange cations have a significant effect on the ion suite of the interstitial waters.

In the muds at the bottom of the Gulf the seawater content of the mud is sufficiently high so that the cations in solution are more abundant than those in the exchange position (based on weight per gram dry clay). At a relatively shallow depth the ratio is reversed, even though the concentration of ions in solution increases. In the Chevron well at 4000 ft, the weight of exchange cations is more than twice the weight of the pore water cations. From 4000 to 11,500 ft, this difference systematically decreases. The exchange ions decrease by 0.3 percent, and the interstitial water ions increase by 0.2 percent. The values for Na are 0.2 percent and 0.2 percent, respectively. Thus the high Na concentration in the interstitial water is not surprising, and it is not entirely the result of selective filtration of seawater.

In the interval from the surface to somewhere above 4000 ft, not only is the amount of ions in the interstitial water less than that in the exchange sites, but the water phase and the permeability decrease. There are not enough free ions so that the exchange sites can be satisfied by the same suite of ions that would be expected if the ion supply was essentially infinite. Ion mobility becomes more important than ion charge, and as demonstrated by Hanshaw (1964) for compacted montmorillonite, the more mobile monovalent cations are adsorbed in preference to the divalent cations. Thus the amount of Na in the exchange positions increases, probably systematically, to a depth of approximately 6000 ft (Fig. 12). Mg is preferentially lost because of its incorporation in the mineral phase and because of the resulting high Ca:Mg ratio in the interstitial water.

At approximately 6000 ft, or deeper, a different selection process becomes operative. Exchange sites systematically become unavailable as illitic and chloritic layers are formed. Cations are released or replaced in the inverse order of the strength with which they are bonded ($\triangle F$). Thus, below 6000 ft Na is removed preferentially from the interlayer position and added to the fluid phase, and the exchange sites are increasingly populated with divalent cations, particularly Ca.

From the surface ($\sim$1.0 percent Na) to 4000 to 5000 ft, the Na content of the sample (in the interstitial water) must be reduced by approximately 0.6 percent (by weight at 110° C). Approximately 0.2 percent or 20 percent of this Na is adsorbed on the clays replacing Mg, 0.4 percent or 40 percent of the Na is presumably flushed from the mud. Through this same interval 1.7 percent Cl (500 epm) is lost. On an equivalent basis, the Na removed accounts for less than half (200 epm) this value. Approximately 0.36 percent Mg would have to be removed to allow a charge balance. Though it is possible that this much Mg is removed, the Na/Cl ratio does not seem probable.

From 4000 to 5000 ft to 9100 ft there is a gradual increase of 0.1 percent of interstitial Na. Between $\sim$ 4000 ft and $\sim$ 7000 ft both interstitial Na and exchange Na increase. Below $\sim$ 7000 ft some exchangeable Na is lost, presumably released to the interstitial water. However, through this interval there appears to be a slight net gain in total Na (interstitial plus exchange). Between 9100 ft and 9800 ft there is an abrupt increase of 0.1 percent Na, to a total of 0.6 percent Na, in the interstitial water. The top of the permeability seal, causing the high-water pressures, apparently occurs at approximately 9800 ft. This sample also has the minimum water content in this part of the section. From the top of the barrier to total depth, the Na content of the interstitial water systematically decreases to 0.35 percent. Below the barrier, the water decreases at approximately the same rate; therefore, the Na concentration remains relatively constant. Approximately 0.2 percent exchangeable Na is also lost through this interval. Thus, there is an apparent decrease of 0.45 percent Na.

It seems reasonable to assume that the deep shales (below 10,000 ft) passed through a sequence of events similar to that of the shallow section. Thus, the Na content in the samples 9800 ft and deeper may have been on the order of 0.3 percent to 0.4 percent, or less, before the permeability barrier was formed. Once the initial barrier formed, the upward migration Na, along with Ca, H_2O, Cl, and lesser amounts of SO_4 and HCO_3 had difficulty passing through it. With increasing depth of burial a gradient was established, and maximum concentration was in the barrier and immediately below it. The release of exchangeable Na could account for almost all the increase in Na between 16,450 ft and 9800 ft. Values for the 10,300-ft sample were so anomalously high that they were originally discarded. In the mechanically formed permeability barrier (due to compaction) the upward migrating Ca formed calcite and increased the effectiveness of the barrier.

The vertical distribution patterns indicate the barrier is most efficient for water. The reversal in gradient of Na between 10,000 ft and 4000 ft (as opposed to that between the surface and 4000 ft and between 10,000 ft and 16,000 ft) suggests there is some migration of Na through the barrier.

Cl behaves somewhat similar to H_2O. Though Cl decreases with depth, there is a relatively abrupt increase ($\sim$ 0.15 percent) between 9100 ft and 10,300 ft, followed by a gradual decrease. Apparently relatively little Cl is getting through the barrier. The concentration of both Na and Cl decreases sharply from $\sim$ 10,000 ft to $\sim$ 12,000 ft (this is particularly evident in graphs of Cl/Al_2O_3, Na/Al_2O_3 values), and then tends to level off. This vertical distribution pattern is, in general, what would be expected when a semipermeable membrane inhibits the movement of upward migrating ions.

The vertical distribution of SO_4 and HCO_3 (increase from surface to top of barrier and then slight and gradual decrease) indicates that they are not very mobile, and the increase in concentrations occurs from the surface to 10,000 ft. Minor amounts, if any, of SO_4 and HCO_3 form below this depth, and once the maximum concentration is formed there is only minor vertical migration.

Most of the available Ca and Mg are lost from the fluid phase above 4000 ft. Most of the Ca leaves the mud system entirely. Some Mg is used to form dolomite in the mud, and some moves into the sands, forming chlorite.

The deepest interstitial waters are dominated by Na and HCO_3 which would have to be the major ions released during the early stages of low-grade metamorphism. It is of interest to note that White (1957) has reported a number of waters with a high Na and HCO_3 content and concluded that they must be metamorphic waters.

In the deeper samples, K and Mg in the interstitial water systematically increase with increasing depth and temperature, presumably due to the increased solubility of the silicate minerals at high temperatures. The bulk chemistry data suggest that K, Mg, and Fe migrate upward to react with K-, Mg-, and Fe-deficient clays to form an illite-chlorite mineral suite. Thus, deeply buried shales are autocannibalistic. The deepest portions of shale sections are exposed to temperatures high enough to cause a significant release of K, Mg, and Fe [granitization (Holmes, 1965)]. These ions migrate upward and react with the K-, Mg-, and Fe-deficient clay minerals to form illite and chlorite. The temperatures in the Chevron wells are not high enough for these latter reactions to be very effective, but the increase in K and Mg in the deeper interstitial waters may be due to this process.

Exchangeable Ca, Ca in the interstitial water, and Ca in calcite (Table 5) are at a maximum in the 10,300-ft sample. This sample also has the highest concentration of Na, Cl, SO_4 (this is the only anion suite dominated by SO_4), and one of the lowest concentrations of HCO_3. These data, as well as the other data, suggest there has been an upward migration of these ions and that Ca has precipitated as calcite, increasing the effectiveness of the permeability barrier. The former physical barrier becomes a chemical barrier.

TABLE 5. PERCENT CALCIUM (CA)

Ft	Nonexchangeable	Exchangeable	Interstitial	Total
4233	1.32	0.69	0.001	2.01
8650	0.31	0.25	0.002	0.56
9100	0.40	0.41	0.006	0.82
9800	0.50	0.44	0.002	0.94
10300	1.52	1.05	0.020	2.59
10530	0.46	0.83	0.003	1.29
11020	0.14	0.21	0.002	0.35
12020	0.23	0.29	0.0003	0.52
16450	0.66	1.03	0.0004	1.69

After the clay was removed from the clay-water slurries, the water from most of the samples had a yellow-to-brown color due to soluble organic material (Table 3). The color tended to increase in darkness with increased depth, with a distinct increase in color in samples below 9100 ft.

The increase in soluble organic material with depth may merely reflect a higher concentration of organic material or it may be due to the increase in temperature and possibly Na concentration. The solubility of humic and fulvic acids in soils and clays, at least partially adsorbed on the clay minerals, is increased by increased temperature and Na concentration, either NaCl or Na_2CO_3 (Swain, 1963). Heat also breaks up the large molecules, increasing the number of functional groups. These groups adsorb ions such as Na, Ca, Mg, Fe, and others (as do porphyrins and soaps). The presence of negatively charged organic material is further suggested by the epm values (Table 4); all deep samples show a deficiency of anions; the deficiency tends to be larger in the samples with the more darkly colored water. The increased conductivity of high-pressured muds may be due to this increase in functional organic material (Harold Overton, oral commun., 1969). Philippi (1965) has estimated that the bulk of the petroleum generation takes place at temperatures above 115° C, although generation apparently starts at appreciably lower temperatures.

The low Ca content of the interstitial waters indicates that it is easily removed from the muds. It may be that some of the Ca remaining is complexed by the soluble organic fraction. The Ca-organic (and Na-organic) complex may migrate upward to where it is trapped by the low permeability muds (80° C). The decrease in temperature and pressure and increase in Na concentration would favor the release of Ca and the formation of calcite or the mechanism may be similar to that which causes the formation of calcite concentrations in organic muds and peats.

Temperature may be as important, or more so, in the release of organic material from clays as it is in the release of interlayer water.

In view of the uncertainties in the interstitial water data, any interpretation must be viewed with skepticism.

Bulk Chemistry

PLIOCENE-MIOCENE

Al_2O_3

The Al_2O_3 content, on the basis of 110° C weight, ranges from 9.94 to 17.47 percent and averages 14.3 percent (Fig. 15). The average value is 1.1 percent less than reported by Clark (1924, p. 34) for the average shale. If it is assumed that the clay suite contains an average of 23 percent Al_2O_3, then most samples contain between 55 and 75 percent clay minerals and feldspar. The feldspar content would be on the order of 5 percent. The electrical resistivity values decrease as the Al_2O_3 increases, as would be expected if the Al_2O_3 were a measure of the clay content. The generalized relation shows that Al_2O_3 systematically increases from 4233 ft to 12,000 to 13,000 ft, and then fairly regularly decreases.

K_2O

The measured K_2O content of the bulk samples (Fig. 15) averages 2.52 percent K_2O and has a depth distribution similar to that for Al_2O_3. Though the Al_2O_3/K_2O ratio (5 to 6) is relatively constant with depth (Fig. 16), there does seem to be a systematic decrease in the Al_2O_3/K_2O ratio (as well as Al_2O_3/MgO and Al_2O_3/Fe_2O_3) below 10,000 ft to 11,000 ft. The environment in which the samples were deposited changes from shelf to bathyal at approximately 9800 ft; the number of deep-water benthonics tests increase markedly below 10,930 ft.

The oxide ratios are relatively constant in the shelf samples and the relatively low values for Al_2O_3/K_2O and Al_2O_3/MgO probably reflect the presence of more detrital feldspar and coarse-grained mica and chlorite in the shelf than in the bathyal samples. The regular increase in K, Mg, and Fe, relative to Al in the bathyal samples indicates a systematic change in the originally deposited detrital mineral suite or that K, Mg, and Fe have migrated into the system from below. Other data suggest the latter interpretation is a reasonable possibility.

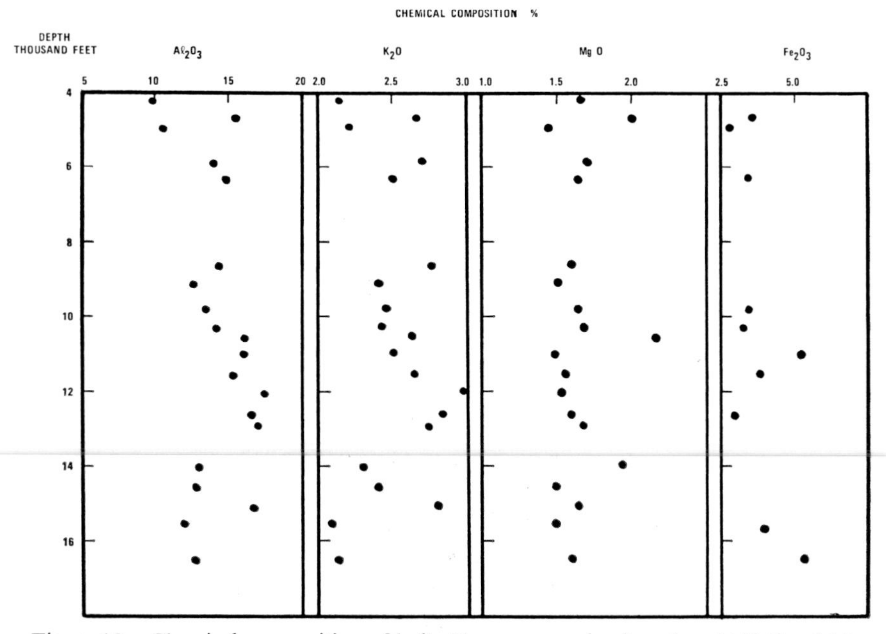

Figure 15. Chemical composition of bulk Chevron samples, based on 110° C weight.

Quantitative X-ray analyses of recent marine mud samples from the vicinity of the Mississippi Delta and Pliocene and Miocene muds from two wells off the coast of Texas (Shaw and Weaver, 1965) indicated that most contained less than 5 percent feldspar (maximum value was 10 percent). In these samples as in the Chevron samples, Na feldspar is more abundant than K feldspar. Na_2O analyses of the bulk material, with the exchange Na removed, indicate that approximately 0.5 percent Na_2O is present. If all the Na were in feldspar, which is unlikely, this would be equivalent to less than 5 percent Na feldspar. The amount of K feldspar would be less.

If as much as 5 percent feldspar is present (chemical analyses of the $> 2\ \mu$ fraction indicate this would be a high value), then 0.75 percent K_2O or approximately 30 percent of the total K_2O could be in feldspar. The actual value is probably closer to half this amount. The amount of K_2O in the clay minerals of the shallow samples is probably closer to 2.0 than 2.5 percent.

In the deeper samples a higher proportion of the K is present in the illite layers. If it is assumed that these layers require 10 percent K_2O to be completely collapsed, then the clay suite should be composed of approximately 40 percent 10 Å layers (2.5 percent K_2O; 60 percent clay minerals). X-ray analyses indicate the mixed-layer clay is composed of approximately equal parts illite-chlorite-montmorillonite; if we include the discrete 10 Å layers the percentage of 10 Å layers in the clay suite would be approximately 37 percent.

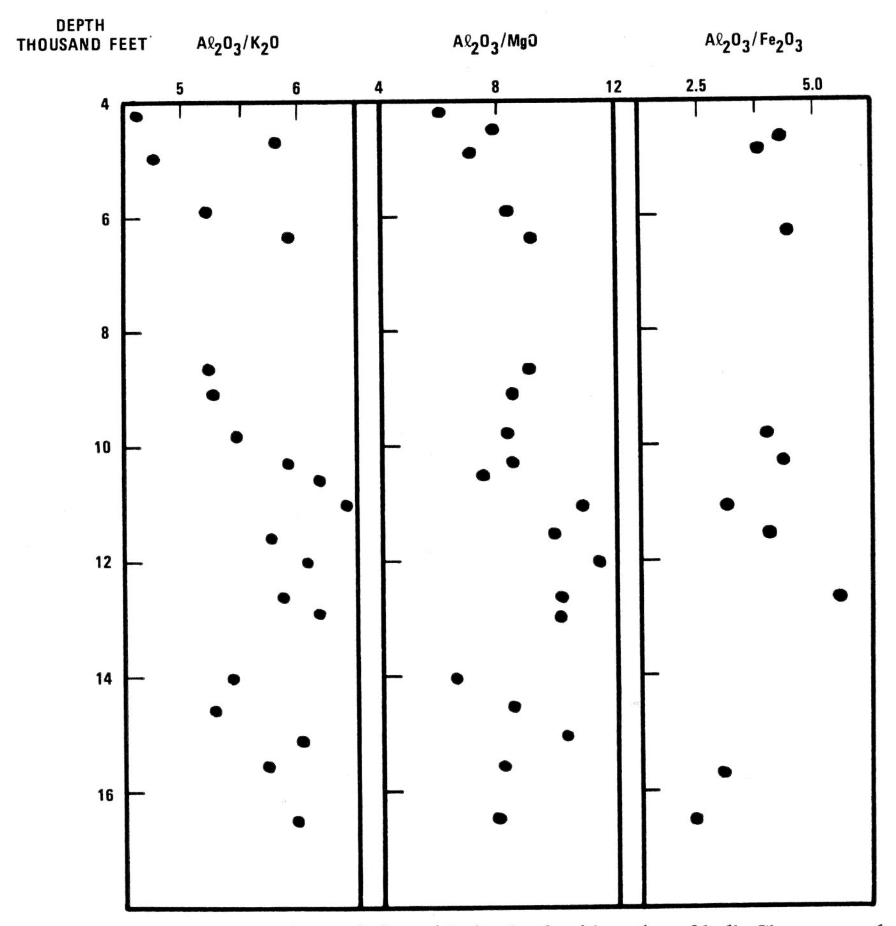

Figure 16. Graphs showing variation with depth of oxide ratios of bulk Chevron mud samples.

If, as is generally assumed, all of the montmorillonitic layers (80 percent of clay suite) were converted to 10 Å mica-like layers during deep burial, the bulk shale should contain approximately 4.8 percent K_2O. If these layers were converted to an illitic material (mixed-layer illite-chlorite) containing only 7 percent K_2O, then 3.4 percent K_2O would be required. Clearly, both the Pliocene-Miocene and Springer material do not have sufficient K_2O to form typical Paleozoic and Precambrian illitic shales.

The presence of kaolinite in the Pliocene-Miocene and Springer montmorillonitic clays explains, in part, the difference between the Al_2O_3/K_2O ratio of these sediments and that of the older illitic shales; however, the difference remains fundamental. Even if the kaolinite is destroyed and the Al goes into or between the montmorillonitic layers, an average of between 0.6 and 2.7 percent K_2O is required to decrease the ratio to that of the older shales. Thus, these clays will not end up as similar to the Paleozoic and Precambrian shales unless K is acquired from an external source. If no K is added, and the temperature becomes sufficiently high, the resulting shale will contain less illite and more dioctahedral chlorite (either interlayered or discrete) than the Paleozoic and Precambrian shales.

From 10,000 to 16,450 ft the proportion of expanded layers decreases and, in general, the amount of clay minerals (Al_2O_3) decreases. At the same time the amount of interstitial K increases. Thus, the amount of K per amount of expanded layers increases drastically in this interval.

There is a linear relation between the percent Al_2O_3 and the Al_2O_3/K_2O ratio of the Chevron samples with the ratio increasing (and the K deficiency increasing) with increasing Al_2O_3 (Fig. 17).

The data shown in Figure 17 indicates that the K deficiency is not restricted to the Gulf Coast Pliocene-Miocene and Springer clays. The Tertiary and Mesozoic shales of the Russian Platform and the recent pelagic muds have Al_2O_3 and K_2O values that lie on or near the trend of the Chevron samples. The Paleozoic and Precambrian shales have significantly larger amounts of K_2O for a given value of Al_2O_3 than the younger sediments. The scattered data suggests that the values for the older shales form a linear trend roughly parallel to that of the younger samples. The low Al_2O_3/K_2O ratios for the samples with a low Al_2O_3 value may be due to the presence of larger amounts of K-feldspar in the sandier samples.

If the trend of Al_2O_3/K_2O values with depth (Fig. 16) is extrapolated, a ratio value similar to that of Precambrian and Paleozoic shales, 4.5, would be encountered at approximately 28,000 ft or a temperature of approximately 210° C (Fig. 22). This is the temperature at which illite is formed.

The average Al_2O_3/K_2O ratio for illites is about 3.7, or larger, than that for most muscovites. Thus, for the amount of K available, there is more than enough Al to form 2:1 layers with the muscovite composition. Phengites and the better crystallized illites have ratios near 2.6 and K_2O values near 10 percent. The Al_2O_3 content ($\sim$26 percent) of these latter samples is the same as for the run-of-the-mill illites.

It appears that, except for K_2O, micas (phengites) can have the same general oxide composition as "poor" illites. The difference in K_2O content and the asymmetry of the 00ℓ peaks indicates the two are not the same. The illite mineral complex, apparently containing both mica and chlorite layers, can presumably be reorganized into a single phase mica of a phengite composition when sufficient K and heat are available. When additional K is not available, discrete units of mica and chlorite would be formed.

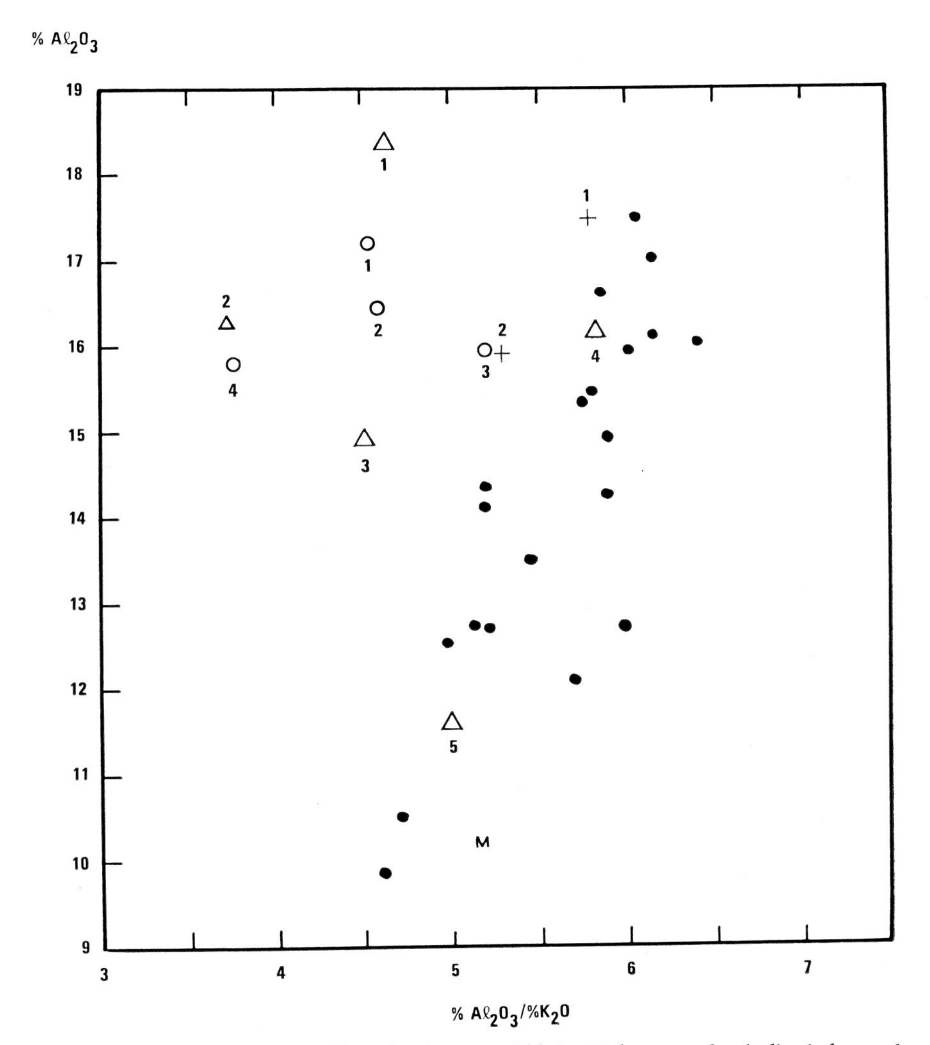

Figure 17. Relation of $\%Al_2O_3/\%K_2O$ to $\%Al_2O_3$. Values are for bulk shales and muds. •, Chevron samples; M, Mississippi River mud; +, Holocene marine muds: 1, Goldberg and Arrhenius (1958); 2, Turekian and Wedepohl (1961). 0, U.S.A.; 1, Precambrian Nanz, 1953); 2, Paleozoic (Clark, 1924); 3, Mesozoic and Cenezoic (Clark, 1924); 4, Ordovician (Scotford, 1965). Δ Russian: 1, Precambrian; 2, lower Paleozoic; 3, middle and upper Paleozoic; 4, Mesozoic; 5, Tertiary (Vinogradov and Renov, 1956).

By using the data of Poldervaart (1955), it can be calculated that the lithosphere contains approximately 1362×10^{15} tons of shale. If all the clay in the shale was 10 Å, 10 percent K_2O illite (assuming a phengite Al_2O_3/K_2O ratio of 2.6 and an Al_2O_3 content of 17 percent for the average shale), 88×10^{15} tons of K_2O would be required. The total K_2O in all sediments is 31.8×10^{15}.

Thus, the availability of K is a definite limiting, but perhaps not controlling, factor in the amount of illite that can form. Approximately 62 $\times$ 10^{15} tons of K_2O would be required for all clay minerals to be a 7-percent K_2O mixed-layer illite-chlorite. Thus, there is enough K_2O in sediments so that about half the clay minerals could be 7-percent K_2O "illite." X-ray studies indicate this is a reasonable value.

Na_2O

Na analyses were made on four samples from which the interstitial and exchangeable Na had been removed. Aside from the shallowest sandy sample the Na_2O and Al_2O_3/Na_2O are relatively constant with depth (Table 6). The Na_2O values are about half those in the average Precambrian and Paleozoic shale (Nanz, 1953), and the Al_2O_3/Na_2O values are approximately twice as large.

When the exchangeable Na is added to the fixed Na the total values are still low relative to the Precambrian and Paleozoic. However, if the Na in the interstitial water is included in the total Na_2O, the Na_2O and Al_2O_3/Na_2O values, particularly in the deepest two samples, are similar to those for the average Precambrian and Paleozoic shale (Na_2O, 1.23 and 1.01 percent, respectively; Al_2O_3/Na_2O, 13.9 and 16.3, respectively).

Although the data is limited, it could suggest that, after a certain depth of burial (temperature higher than 100° C?), little exchangeable and interstitial Na is lost from the system and with increasing temperature becomes incorporated in the solid phase, presumably as albite, thus making the low-grade metamorphic rock more sodic than its sedimentary precursor.

MgO

Relatively, the MgO content of the Chevron samples (average 1.66 percent) remains uniform with depth (Fig. 15). The trend of the Al_2O_3/MgO values (Fig. 16) is similar to that of Al_2O_3/K_2O.

TABLE 6. NA DISTRIBUTION IN CHEVRON SAMPLES

Depth (ft)	Fixed Na_2O (percent)	$\frac{Na_2O}{Al_2O_3}$	Fixed and Exchange-able Na_2O (percent)	F + E* $\frac{Na_2O}{Al_2O_3}$	Total (percent)	Total $\frac{Na_2O}{Al_2O_3}$
4233	1.17	8.5	1.40	7.1	1.68	5.9
8650	0.524	24.4	0.86	16.7	1.35	10.6
12020	0.533	32.8	0.78	22.4	1.22	14.3
16450	0.393	32.5	0.49	26.0	0.88	14.6

* Fixed and exchangeable

Apparently the uniformity of the MgO values is due mostly to the fact that Mg occurs both in dolomite and in the clay minerals. The shallow and deep samples contain less clay than the middle samples, but contain more dolomite (X-ray). If the clay minerals in the sandier samples have the same Al_2O_3/MgO ratio (10) as those in the clay-rich samples, approximately 2 percent dolomite should be present. This is a reasonable figure, but would be slightly low because of trace amounts of dolomite in the clay-rich samples. Wickman's (1954) calculations indicate that the average shale (*from* Clark, 1924) contains slightly less than 2 percent dolomite.

For a given value of Al_2O_3, the Al_2O_3/MgO ratio of the Tertiary muds (including the Tertiary muds of the Russian Platform) is less than that of the Paleozoic and Precambrian shales (Fig. 18). The difference in MgO content is approximately 1 percent. A significant difference between the Mg and K data is that the recent marine (largely pelagic) averages have lower Al_2O_3/MgO ratios per amount of Al_2O_3 than any of the other samples. Presumably, this difference is due to the large amounts of montmorillonite forming from volcanic material in the present-day pelagic environment or the longer time the material is in contact with seawater, or both. The high Al_2O_3/MgO ratio of the Mississippi River sample suggests that some Mg has been acquired from the seawater, either before or after burial, or both.

The average illite contains 2.75 percent MgO. If it is assumed that Precambrian and Paleozoic shales contain an average of 75 percent illite, then 2.06 percent MgO is present in the illite component of the shales. These values indicate the chlorite component contains only 2.8 percent MgO. Illites formed from the weathering of feldspars contain less MgO—on the order of 1 percent (Weaver, 1971). As most ancient illites were formed from feldspar, this is a more reasonable value. However, even this value allows only 8 percent MgO to the chlorite. These calculations indicate that either the amount of chlorite (25 percent) is overestimated or the sedimentary chlorites are Fe or Al-rich, or both. The latter supposition is supported by the X-ray data. Little or nothing is known about the composition of chlorite in shales. The Tertiary muds will presumably form shales even richer in Al-chlorite than the Paleozoic-Precambrian shales.

Fe_2O_3 and FeO

The iron content (average 3.8 percent as Fe_2O_3) of the Chevron samples (Table 7) is approximately half that for the averages for Paleozoic and Precambrian shales. As with K_2O and MgO, the Al_2O_3/Fe_2O_3 ratio increases as Al_2O_3 increases and the ratio values (as well as the Russian Tertiary) are higher than those of the older shales (Fig. 19). The recent marine mud values lie with those of the older shales. Fe-rich clay minerals appear to be the only ones forming in the present marine environment.

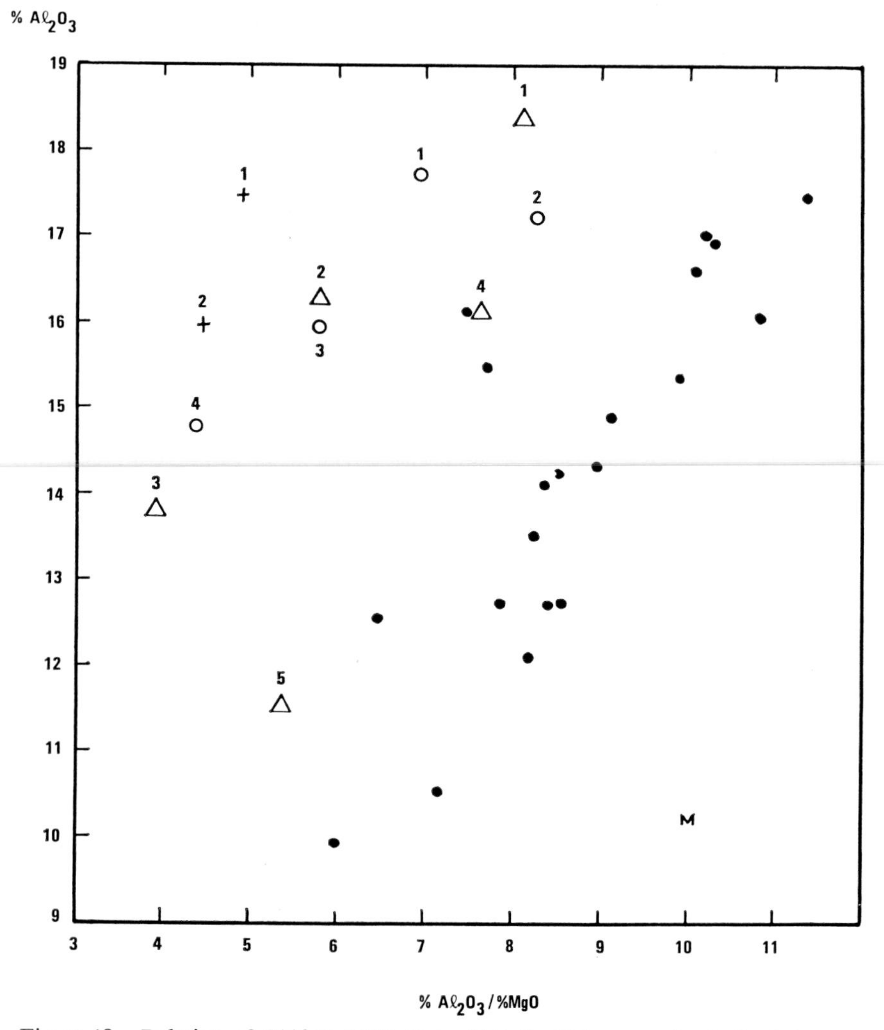

Figure 18. Relation of $\%Al_2O_3/\%MgO$ to $\%Al_2O_3$. Values are for bulk shales and muds. Significance of symbols is the same as for Figure 17.

Total iron shows little variation with depth, though two samples have significantly higher amounts of iron. Seven samples were treated to remove free iron oxide (Mitchell and Mackenzie, 1954). From 14 to 65 percent of the iron (as Fe_2O_3) was readily soluble. Most values were near 25 to 30 percent. Most of the soluble iron was in the form of Fe_2O_3. There was no trend with depth.

The amount of nonsoluble Fe_2O_3 is relatively constant with depth; however, the amount of FeO is much larger (and the Al_2O_3/FeO much lower) in the

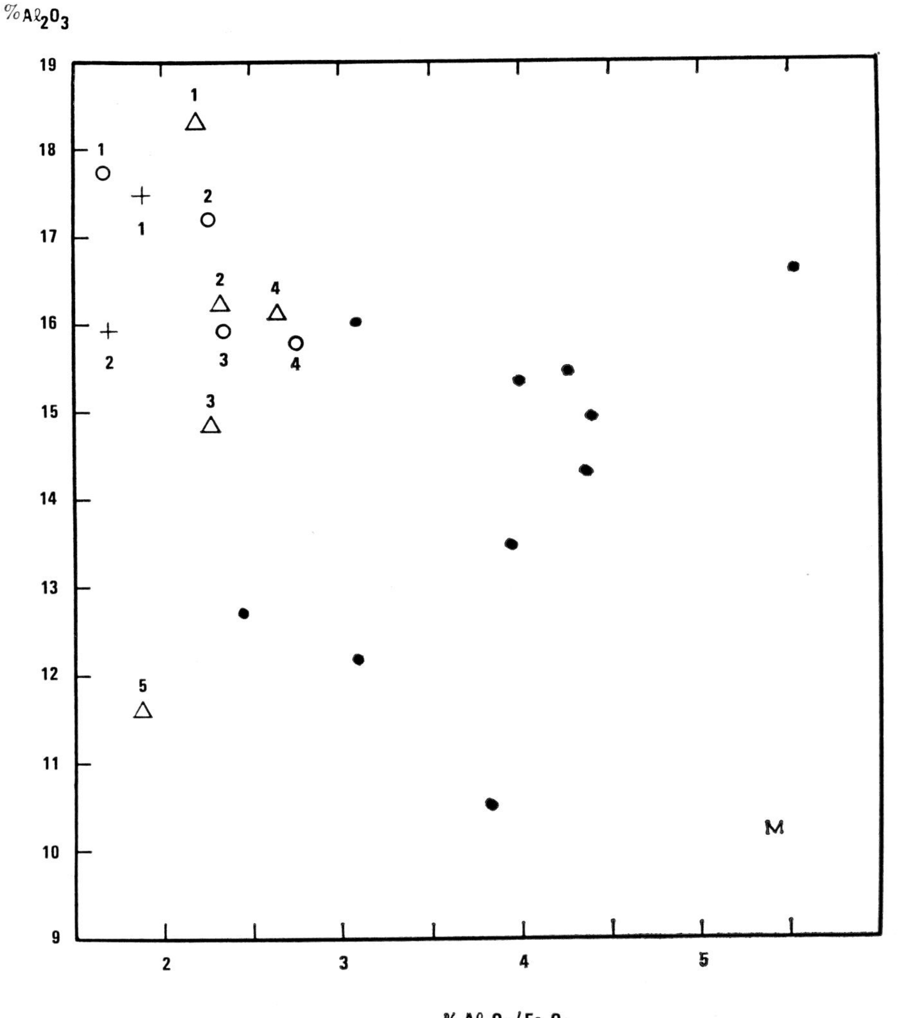

Figure 19. Relation of $\%Al_2O_3/\%Fe_2O_3$ to $\%Al_2O_3$. Values are for bulk shales and muds. Significance of symbols is the same as for Figure 17.

deepest of samples. The FeO/Fe_2O_3 ratio of the "non-free" iron is larger than one in all samples (Table 7). Though the data is somewhat limited, it does suggest that the deeper samples have a larger ratio. Mossbauer analyses indicate that the Fe^{2+} is relatively more abundant in the deeper samples and that the Fe is present in several different environments. The spectra are so complex that specific assignments of the Fe are difficult.

The FeO/Fe_2O_3 value for the averaged Precambrian shale analyses is larger

TABLE 7. IRON VALUES BEFORE AND AFTER REMOVAL OF "FREE" IRON

	Analysis of Untreated Sample				After Removal of "Free Fe"		
	Depth in Feet	FeO	Fe as Fe_2O_3	Fe_2O_3	FeO	Fe as Fe_2O_3	Fe_2O_3
(2)	4710	2.37	3.63	2.56	2.82	3.11	1.84
(3)	4950		2.76				
(5)	6350	1.43	3.41, 3.37	2.77, 2.73	1.59	2.26	1.54
(8)	9800		3.42				
(9)	10300	2.14	3.27	2.31	2.58	2.42	1.26
(11)	11020		5.20				
(12)	11520	2.18	3.83	2.85	1.84	1.34	0.51
(14)	12610	1.97	2.99	2.11	2.00	2.27	1.37
(20)	15650	5.51	3.90	1.42	5.01	2.78	0.53
(21)	16450	4.24	5.22	3.31	3.91	3.76	2.01

than one; for Paleozoic and younger (including present-day marine) the ratio is less than one (Nanz, 1953, p. 60). The shallow Chevron samples (total Fe) have ratios slightly less than one, but, in general, larger than for the Paleozoic and younger shales. The 15,650- and 16,450-ft samples have values larger than one.

If it is assumed that all of the "non-soluble" FeO is present in the alumino-silicate clay minerals, then the concentrations would be from about 3 to 10 percent FeO. Illite and montmorillonite commonly contain less than 0.5 percent FeO (illite averages 0.6 percent, montmorillonite averages 0.2 percent). Relatively few glauconites contain more than 3 percent FeO. All of these clay minerals have Fe_2O_3/FeO ratios from 5 to 20. Thus, it is likely that much of the Fe^{2+} is present in the chlorite and chloritic layers, although some is present in siderite, particularly in the deeper samples.

Shaw (1956) and others have shown that the Fe^{2+}/Fe^{3+} values increase during low-grade metamorphism. In the Salton Sea geothermal field (Muffler and White, 1969) the ratio increases with increasing temperature. The ratio exceeds one at a temperature somewhere between 120° and 150° C. Most of Fe^{2+} is used to form chlorite. It would appear from the Chevron samples that when considerable organic material is present, reduction can occur at lower temperatures and pressures, and presumably without the need of sulfate-reducing bacteria and the formation of pyrite.

It is interesting that the Al_2O_3/Fe_2O_3 (total) values of the Salton Sea samples are similar to the Chevron samples, except for the deepest sample (~235° C), even though the area is permeated with an iron-rich brine.

The total Fe_2O_3 content of the samples shallower than 10,300 ft averages 3.3 percent. The deeper samples average 4.2 percent. The difference is due mostly to an increase in the amount of FeO in the deeper samples. This change approximately coincides with a change in depositional environment from

TABLE 8. BULK CHEMISTRY OF SPRINGER SHALES

Sample	Al_2O_3	K_2O	Na_2O	MgO	CaO	Fe_2O_3	$\frac{Al_2O_3}{K_2O}$	$\frac{Al_2O_3}{MgO}$	$\frac{Al_2O_3}{Fe_2O_3}$
(1) outcrop	16.84	1.68	0.2±	1.49	0.09	3.68	10.0	11.3	4.6
(2) outcrop	23.12	1.56	0.25	2.22	0.04	4.13	14.8	10.4	5.6
(3) 20740–50	17.96	2.03	0.72	1.70	0.48	3.58	8.8	10.5	5.0
(4) 21870–50	21.66	2.02	0.78	1.76	0.25	3.44	10.7	12.3	6.3
(5) 21960–70	23.94	2.00	0.85	1.84	0.05	4.35	12.0	13.0	5.5
(6) 23190–200	25.36	2.25	1.10	1.73	0.17	3.33	11.3	14.6	7.6
(7) 23510–20	25.14	2.72	0.88	1.82	0.11	3.35	9.2	13.8	7.5
(8) 23610–20	23.98	2.77	0.83	2.03	0.08	3.90	8.6	11.8	6.1
Average	22.25	2.13	0.70	1.81	0.16	3.72	10.7	12.3	6.0

neritic to bathyal, which may account for the increase in total iron, or the increase may be due to the addition of iron from deeper in the section. Detectable (by X-ray) siderite occurs in the samples below 15,500 ft.

MISSISSIPPIAN

The Springer Shale has a higher clay mineral content, as indicated by the Al_2O_3 values (Table 8), than the Chevron and average Paleozoic and Precambrian shales. The relatively high kaolinite content is partly the cause of the high Al_2O_3 values. The K_2O, MgO and Fe_2O_3 values are all less than those for the Paleozoic and Precambrian shales, even though the average Al_2O_3 value is larger.

The plot of Al_2O_3 *versus* Al_2O_3/K_2O values is scattered. The deepest two samples fall on trend with the Chevron samples, but the other samples are even more deficient in K_2O than the Chevron samples.

The Al_2O_3/Fe_2O_3 values fall on trend with the Chevron samples, larger per amount of Al_2O_3 than the Paleozoic and Precambrian samples. Except for two shallow samples, the Al_2O_3/MgO values are on trend with the Paleozoic and Precambrian values, though the ratio values are considerably larger. This reflects the relatively high chlorite content of the Springer Shale compared to the Pliocene-Miocene, and possibly indicates an increased availability of Mg in Paleozoic time, either in the source material or in the seawater. The other possibility is that by 21,870 ft (depth of shallowest sample to fall on trend of older shales), Mg has been added from fluids originating from below.

The K_2O values of the eight Springer Shales average 2.13 percent (10 other samples from scattered depths averaged 2.3 percent). The deepest two samples (23,510 and 23,610 ft) average 2.75 percent as contrasted to values of near 2.0 or less for shallower samples. This could mean that some K has been added to the system.

SUMMARY

K, Mg, and Fe are less abundant both in absolute values and relative to Al, in the montmorillonitic muds and shales (Chevron and Springer) than in the Precambrian and Paleozoic shales. All three ions, relative to Al, tend to increase with depth. This increase may be due to changes in the detrital suite or the addition of ions from deeper in the section. If it is assumed that the nonkaolinite components of the Chevron samples have the same Al_2O_3/K_2O as in the older shales, then the Chevron samples with 16 percent Al_2O_3 contain an excess of approximately 4 percent Al_2O_3; for the Al_2O_3/MgO ratio an excess of 6 percent Al_2O_3 is obtained; for Al_2O_3/Fe_2O_3 the excess Al_2O_3 is 8 percent. The excess Al_2O_3 would be larger for the Springer samples.

These values, 4 to 8 percent, are equivalent to 10 to 20 percent kaolinite (which is seldom present in the early Paleozoic and Precambrian shales). This is equivalent to approximately 15 to 40 percent kaolinite in the clay-mineral fraction. These values would seem to be high for both the Chevron and Springer samples. It seems likely that some of the chlorite and chloritic interlayers are Al-rich and partially dioctahedral. When the kaolinite that is present is destroyed, and the Al is redeployed, the proportion of dioctahedral chlorite layers should increase further.

It appears that both young muds and ancient shales have appreciable dioctahedral Al-rich chlorite and the younger material will eventually have a higher proportion unless ions are added from outside the system.

PREVIOUS STUDIES

Burst (1969), Kossovskaya, (1960) and Dunoyer De Segonzac (1964) have described the conversion of montmorillonite, by way of mixed-layer illite-montmorillonite, to illite with increasing depth of burial. Fuchtbauer and Goldschmidt (1963), Dunoyer De Segonzac (1964), and others (*see* Muller, 1967, for review) have noted a decrease in the kaolinite/chlorite ratio with depth and suggested the destruction of kaolinite and the formation of chlorite.

The senior author has made X-ray studies of hundreds of oil wells in North America and though he has seen these trends, he has never seen them go to completion—the conversion of montmorillonite to 10 Å illite and the complete destruction of kaolinite. Most of the published X-ray patterns do not indicate "completion."

In the Ouachita Mountains of Texas, a region of low-grade metamorphism, kaolinite is destroyed and its loss is the first effect of metamorphism. As low-grade metamorphism increases, mixed-layer illite-montmorillonite is converted to mixed-layer illite-chlorite, and the narrowness of the illite 10 Å peak decreases progressively.

This latter trend in areas of low-grade metamorphism has been substantiated by studies of Kubler (1964) and Dunoyer De Segonzac and Kubler (1968).

That a montmorillonite, illite, kaolinite, chlorite (in order of decreasing abundance) clay suite can be converted to an illite-chlorite clay suite seems certain. The conditions and mechanisms are in question.

Muffler and White (1969) have been able to relate the effects of temperature on clay minerals by studying sediments from the Salton Sea geothermal system. The Salton Sea area has active hydrothermal metamorphism. Discrete montmorillonite disappeared at temperatures less than 100° C. The resulting mixed-layer illite-montmorillonite shows a decrease in the proportion of expanding layers with increasing temperature (and depth). This material is converted completely to a 10 Å K-mica (it is not known whether interlayers of chlorite are present) at temperatures of 210° C or slightly lower.

The amount of kaolinite decreases and the amount of chlorite increases with increasing temperature. Kaolinite begins to decrease and chlorite begins to increase at approximately 165° C. Kaolinite persists to a maximum temperature of 200° C, though most of it is gone at temperatures slightly higher than 165° C. In one well, chlorite started to form at a temperature of approximately 130° C.

The chlorite is trioctahedral, and the brucite layers are composed predominantly of Mg. It is believed to have formed from kaolinite, dolomite, and ankerite with some Fe added from the brine.

Steiner's (1967) study of the Wairakei thermal area, New Zealand, showed that montmorillonite alters to mixed-layer illite-montmorillonite at 120° to 130° C; the amount of illite increases with increasing temperature. K is supplied by the geothermal fluids. The mixed-layer clay is not completely converted to 10 Å illite until the temperature is higher than 230° C. In this latter stage, illite comprises only 25 percent of the sample (compared to 100 percent for the mixed-layer clay). The accompanying clay is presumably chlorite. Fe-chlorites start to form at temperatures of less than 100° C and are present up to the maximum temperature of 265° C. Both the illite and chlorite are formed largely from the alteration of montmorillonite.

The Al_2O_3/MgO values of the samples from the Salton Sea area do not show any systematic change with depth, which presumably is due to the relatively low Mg content of the brines (Muffler and White, 1969, p. 161).

A plot of the Al_2O_3/MgO ratio *versus* depth of the data of Dunoyer De Segonzac (1964, p. 298) gives a better picture (Fig. 20). In a series of core samples (surface to 4000 m) from the Tertiary of the Cameroun, Africa, he observed a mineralogic change similar to that found in the sediments already discussed; that is, with depth, kaolinite decreases, and chlorite increases;

montmorillonite is converted to a mixed-layer illite-montmorillonite, and at depth (4000 m), illite (though the deepest sample has a 10.5 Å "illite" peak and is apparently a mixed-layer illite-chlorite-montmorillonite).

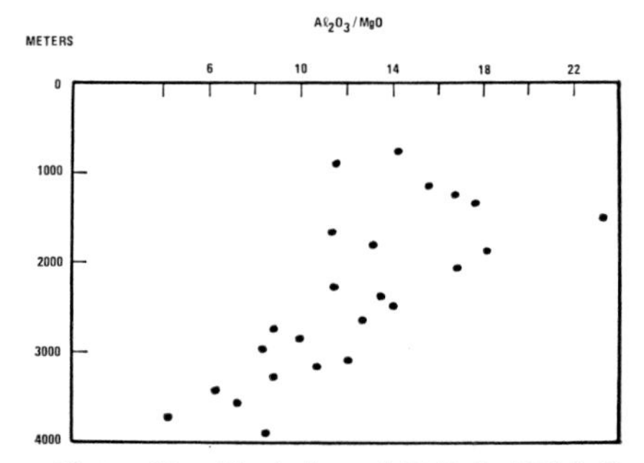

Figure 20. Variation of $\%Al_2O_3/\%MgO$ values of bulk shale samples with depth. Tertiary samples from Cameroun, Africa (Dunoyer De Segonzac, 1964).

Though there is some scatter, probably due to variations in the original composition, the Al_2O_3/MgO ratio systematically decreases with increasing depth (Fig. 20). The Al_2O_3/MgO *versus* Al_2O_3 plot indicates that the shallower samples lie on trend with the Chevron samples; deeper samples with higher chlorite content have a trend similar to the Paleozoic and Precambrian shales. These data suggest that external Mg has been added. These mineral and chemical changes take place at a much shallower depth than equivalent changes in the Tertiary and Springer shales of this study and suggest a much higher geothermal gradient.

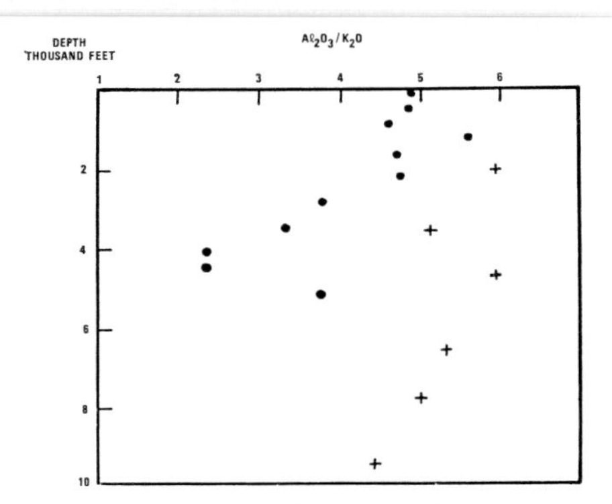

Figure 21. Variation of $\%Al_2O_3/\%K_2O$ values of bulk shale samples with depth. Samples from two wells in the Salton Sea geothermal area (Muffler and White, 1969). Dots represent data from I.I.D. No. 1 and crosses from Wilson No. 1.

The Al_2O_3/K_2O *versus* depth plot of the Muffler and White (1969) data shows a systematic decrease with increasing depth (Fig. 21). The plot of Al_2O_3/K_2O *versus* Al_2O_3 shows that as the Al_2O_3 content increases, the ratio decreases. These plots indicate K has been added to the system. The deeper samples that fall on the trend of the Paleozoic and Precambrian samples have temperatures higher than 260° C. Samples exposed to temperatures between approximately 160° C and 260° C occur between the two trends, and lower temperature samples lie on the Chevron trend. In part, this trend also reflects an increase in K-feldspar with increasing depth. Unfortunately, Dunoyer De Segonzac (1964) does not give any K values.

These data indicate that not only the mineralogy, but also the Al_2O_3/MgO and Al_2O_3/K_2O values of muds and shales increasingly resemble those of the Paleozoic and Precambrian shales as the temperatures to which they are exposed increase. Shales resembling these older shales apparently can form from montmorillonite-rich muds at temperatures of 150° to 230° C. K, Mg, and Fe must be supplied from external sources if the change is to go to completion (10 Å illite and trioctahedral chlorite greater than dioctahedral chlorite).

By the same reasoning, if the Pliocene-Miocene muds and Springer Shale are to resemble the Precambrian and Paleozoic shales, they must be buried deeply enough so that temperatures are high enough to cause the destruction of kaolinite, the movement of Al into the tetrahedral layer, and the invasion of K, Mg, and Fe, possibly, but not necessarily, as hydrothermal fluids. If these three cations are not added, the resulting clay should have a relatively high proportion of dioctahedral chlorite, and much of it occurs interlayered with the illite. The resulting illite would be a mixed-layer illite-chlorite containing from 4 to 6 percent K_2O. Muffler and White's (1969) study of regional hydrothermal metamorphism in the Salton Sea geothermal system suggests that a general soaking has occurred in this area and that the movement of solution in parts of the area "... is probably not much greater than in ordinary sedimentary basins and metamorphic environments."

In the Ouachita area, hydrothermal conditions are presumably not intense, but veins containing abundant chlorite (trioctahedral) are "profuse" and increase in abundance with increasing degree of metamorphism (Flawn and others, 1961, p. 118). Thus, sufficient Mg and Fe and, presumably, K could have been supplied by hydrothermal fluids to react with the Al-rich detrital clay and produce the characteristic illite-chlorite mineral assemblage and composition.

It should be mentioned that the vertical and lateral migration of these cations through a mud or shale sequence need not require the migration of the fluid phase.

A regional vertical migration of ions, such as would be required in the Gulf Coast, sufficient to alter a thick mud sequence to an illite-chlorite shale (Paleozoic type) could represent the feather edge of granitization. These ions could be the outgoing "emanations" from rock that is being transformed into granite. Holmes (1965, p. 185) indicates Mg, Fe, and K are the most common elements expelled during granitization.

HEAT FLOW

Jones (1968) discusses the heat flow conditions in the northern Gulf of Mexico basin. In this area the geothermal-gradient map conforms with the structural map, with the steepness of the gradient varying inversely with the thickness of unconsolidated sediments. Jones states:

If the geothermal flux is uniform over broad areas, then the thermal conductivity of rock must vary inversely with the geothermal gradient. Sediments that overlie the deepest parts of the Gulf Coast geosyncline would, then, appear to possess the highest thermal conductivity. If they do not, they must form a thermal sink and are now storing heat energy received from below; their temperature must inevitably rise.

However, as the northern Gulf of Mexico is fringed by the partially buried Ouachita-Appalachian Mountain chain, it cannot be assumed that heat-flow values are uniform in this area.

Jones believes there is reduced geothermal flux above and an overheating of sediments below the high-pressured reservoirs, which is due to dehydration of montmorillonite (an endothermic process) and a resulting increase in porosity. Another factor is the decreased upward flow of water caused by the low-permeability muds.

Burst (1969) also relates the dehydration of montmorillonite to the geothermal gradient. He believes that all but the next-to-last layer of water is removed at a temperature of approximately 221° F and that the final layers are removed only at a considerably higher temperature. His interpretation is based on the observation that the first occurrence of mixed-layer illite-montmorillonite is restricted to a temperature range of 182° to 232° F. The fact that some 10 Å illitic layers form from montmorillonite (to form the mixed-layer phase) demonstrates that all the layers of interlayer water can be removed in the 182° to 232° F temperature range (and lower); however, the fact that expanded clays with interlayer water exists at much greater depths and temperatures indicates that temperature is not the controlling factor. Composition is, presumably.

In the Chevron well, the temperature increases from 39° C at 4000 ft to 119° C at 16,500 ft. These values are relatively low for this general area. Jones (1968) reports that formation water is commonly at temperatures greater than 121° C at depths ranging from 10,000 to 14,000 ft and lists temperatures as high as 185° C, which may explain why diagenesis has not proceeded as far in the Chevron well as in wells to the northwest of the South Pass area.

The limited data from the Chevron well (Fig. 22) suggest that the geothermal gradient from 4000 ft to 9600 ft is similar to that between 13,000 ft and 16,500 ft. From 9600 ft to 13,500 ft, which includes the top of the high-pressure interval, the gradient is considerably steeper. This may be due to the use of heat for the dehydration of montmorillonite (heat sink) or may reflect the increased water content (conductivity) of the muds in this interval, or both (Fig. 8). The vertical temperature variation parallels the water variation.

Data from the Anadarko basin indicate the geothermal gradient is similar to that for the Chevron well. Levorsen (1967) reports a bottom hole temperature of 123° C at a depth of 17,823 ft in a well from Caddo County. In the deep well described in this report, the temperature may be in excess of 150° C (probably ~ 170° C) at ~ 24,000 ft.

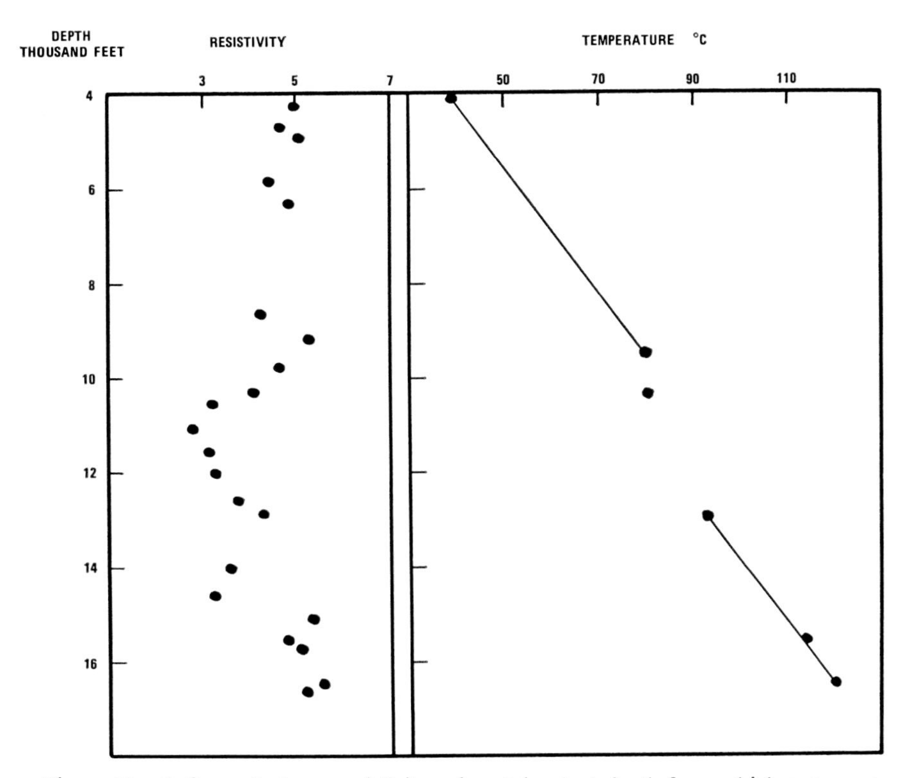

Figure 22. Left graph shows resistivity values (ohms) at depth from which cores were obtained in the Chevron well. Right graph shows measured temperatures for Chevron well.

In the Chevron well, montmorillonite starts to convert to a mixed-layer illite-montmorillonite at a temperature of approximately 50° C, though the 17 Å peak of the glycolated samples is not lost until 80° to 90° C. In the Salton Sea area (Muffler and White, 1969), this latter change occurs below 100° C and in the New Zealand area (Steiner, 1967) at 120° to 130° C. Compositional variations may be largely responsible for these differences. In the Springer well, "complete" conversion of montmorillonite to illite (mixed-layer illite-chlorite) occurs at approximately 170° C. In the Salton Sea, this occurs at 210° C and in New Zealand, at 230° C.

Kaolinite is not destroyed at 119° C in the Chevron well, but is probably destroyed at a temperature of less than 150° C in the Springer well. In the Salton Sea, kaolinite starts to be destroyed at approximately 165° C and does not persist at temperatures above 200° C.

In New Zealand, Fe-chlorites form at temperatures below 100° C; in the Salton Sea, Mg-chlorites start to form at temperatures ranging from 130° to

165° C. In the Springer well, discrete Al-chlorite apparently is formed at a temperature of approximately 150° C.

The diagenesis (formation of discrete chlorite and illite) reported from Africa (Dunoyer De Segonzac, 1964) and Russia (Kossovskaya, 1960) at relatively shallow depths suggests that in these areas the geothermal gradient and heat flow values are greater than in the areas studied in this investigation. Heat flow values given by Lee (1965) suggest this is probably the case. Kossovskaya and Shutov (1955) report that in some Russian basins deep-seated epigenesis occurs at pressures above 1500 atm and temperatures between 150° and 200° C.

This also raises the possibility that the illite-chlorite clay suite of the older Paleozoic and Precambrian shales may be due to higher rates of heat flow and that the clay minerals might be used to supply information on variation in heat flow through geologic time.

Heat flow may have been high enough in the Precambrian and lower Paleozoic to convert relatively shallow buried clays to the stable illite-chlorite clay suite. The "mid-Carboniferous continental drift episode" (Irving, 1966), may have been accompanied by a general decrease in heat flow resulting in a decrease in the amount of illite and chlorite in younger sediments (Weaver, 1967b). The continents would presumably drift from areas of high heat flow to areas of low heat flow. Although it is generally assumed that if the montmorillonitic clays are buried deeply enough, they will alter to Paleozoic-type shales, the data on the Tertiary and Springer (Upper Mississippian) clays indicate this transition is not possible unless significant amounts of K are added. Otherwise, it is likely the conditions under which pre-Pennsylvanian clays formed were significantly different from those in which the younger clays were formed.

Another possibility is that, until the continents started to break up, there must have been large, relatively shallow inland seas with limited connection with the deep oceans. It is conceivable that under these conditions the K, Mg, and Fe content of such inland seas could become higher than in the open oceans and that illitic and chloritic clays would form more readily than in the open oceans which presumably had a composition similar to the present-day oceans. The additional possibility that the heat flow was greater would emphasize further the difference between pre-Pennsylvanian clay suites and the more recent ones. The authors favor the idea that the K, Mg, and Fe has been added after burial and from within the crust.

Compaction

A number of people (Jones, 1968; Dickey and others, 1968; Powers, 1967; Dickinson, 1953) have discussed the occurrence of abnormally high pressures in the sediments of the northern Gulf of Mexico basin. Pressures much higher than normal hydrostatic pressure are encountered at depths ranging from 7000 to more than 16,000 ft. The high pressures are reported for sands, but presumably exist in the surrounding shales which are more porous than normal shales at these depths.

Kerr and Barrington (1961) and Myers and Van Siclen (1964) showed that in many wells in the Gulf Coast, although shale density increases with depth, there is an abrupt decrease in shale density when high-pressure zones are encountered. This decrease is followed by a further systematic increase in density. This phenomenon is seen in the Chevron well by the systematic decrease in the water content of the clays to 10,300 ft where high pressures are first encountered. At this depth there is an abrupt increase in the water content of the muds, followed by a systematic decrease (Fig. 8).

In the Chevron well and many of the other wells in the northern Gulf Coast, the upper portion contains a high percentage of sand and the lower portion contains relatively little. The high-pressure zones start at the top of the thick shale section (Dickinson, 1953). Another pertinent observation is that the water in the sands of geopressured reservoirs decreases in salinity with depth (Jones, 1968; Myers and Van Siclen, 1964). It is not clear whether this trend is restricted to the high-pressure intervals; apparently, it is not.

All of the authors mentioned above conclude that the high pressures are a result of a permeability seal, and most authors suggest that the release of interlayer water from between the montmorillonite layers is an important factor.

The first general observation is that, although montmorillonite is losing its interlayer water throughout the whole Gulf Coast, all deep wells do not have high-pressure aquifers. Pressures may increase suddenly or gradually (Dickinson, 1953). In the Springer Formation of Oklahoma, the Lewis Shale of the Washakie basin, Montana, parts of California, and numerous other areas, the same modification of montmorillonite has occurred, but the areas generally are not characterized by abnormally high pressures.

Data presented by Jones (1968) shows that the geostatic ratio (ratio of fluid pressure to overburden load) at similar depths in the Gulf Coast varies from 0.92 to 0.51. Dickey and others (1968) showed that in a well a high fluid pressure may exist, and in an adjacent well, it may not. They believed this phenomenon was due to faults acting as a barrier to the lateral escape of the water.

These data indicate that the release of interlayer water is not the controlling factor, but most probably a permeability seal is; however, the release of interlayer water may be a necessary factor.

Even when montmorillonite is dehydrated at temperatures of 400° to 500° C, it will rehydrate quite easily. DTA studies by Stone and Rowland (1955) showed that the peaks due to interlayer dehydration shifted from 130° to 175° C at 1 atm to 225° to 265° C at 6 atm. Khitarov and Pugin (1966) found that montmorillonite contained two water layers at 150° C and 5000 atm, and one water layer at 200° C and 10,000 atm. Their laboratory data indicate that under the geothermal conditions existing in the area of the Chevron and Oklahoma wells (~20° C/km), montmorillonite should dehydrate at a depth of approximately 20,000 ft. However, under the conditions of their experiments, it appears that all of the pore water was removed. In the presence of pore water one might expect that higher temperatures would be required for dehydration. The studies in the Salton Sea and New Zealand geothermal fields and the present study, as well as others, indicate that mixed-layer illite-montmorillonites persist to temperatures of 210° to 230° C.

The fact that expanded montmorillonite layers maintain their swelling ability to temperatures of 210° to 230° C, and presumably even as high as 400° C when the composition is right (or wrong), indicates that at temperatures below 210° to 230° C, temperature is probably not so important as composition in removing the final two layers of water. A few montmorillonite layers contract at a certain depth and temperature to 10 Å and from a mixed-layer clay, which probably means that dehydration due to temperature has started; however, expanded layers still persist, no matter how few, at much higher temperatures and for long periods of time (325 m.y. for Springer), which suggest dehydration due to temperature alone, at temperatures less than 210° to 230° C, is unlikely. Even dehydration at temperatures higher than this presumably necessitates some composition change. Dehydration at temperatures less than ~200° C must be due primarily to chemical changes, which are accelerated presumably by temperature increases. Suggested mechanisms for chemical and structural changes have been discovered elsewhere.

The type of dehydration discussed above is permanent dehydration. It is possible that at a certain depth and pressure, montmorillonites exist in a dehydrated state. When the pressure and temperature are lowered, by bringing samples to the surface, the dehydrated layers readsorb pore water and expand. It seems likely, even under relatively low permeability conditions, that it would be easier to remove pore water than the last few layers of interlayer water by increasing pressure and temperature. This is suggested by the experimental data. Engelhardt and Gaida (1963) applied confining pressures of up to 48,000 psi on montmorillonite and reduced the porosity from 80 to 20 percent and found no change in the (001) spacing.

Powers (1967) quoting Martin (1960) assumed the density of the last few layers of interlayer water is 1.4, and that the release of this water to the pores causes an increase in volume and pressure. Burst (1969) calculated that the last water layer has a density of less than 1.00, but other layers may have a larger density. Martin (1960) in his review states, "Unfortunately one may reasonably question virtually all the data cited for various reasons; hence the nature of adsorbed water remains unknown." Without giving a detailed discussion of the literature, it can be stated that the density of interlayer water has not been determined with any degree of certainty; furthermore, it is unlikely that it is of any quantative importance in the development of high pressures.

It should be pointed out that probably the density and amount of interlayer water is variable. The lateral completeness of the water network is difficult to establish. Martin (1960) notes that, when there is not enough water per gram of montmorillonite to form one complete layer, the X-ray data indicate two layers are present. Large cations are known to disrupt the water structure; smaller cations have little effect. Davidtz and Low (1968) concluded that the amount of swelling and the O-H bond length (and thus the density) are related to the amount of rotation of the silica tetrahedrons.

Powers (1967) states that at "1,500–3,000 feet, most of the water is water of hydration which is stacked at least four mono-molecular layers thick between the unit layers of montmorillonite." Burst (1969) states that after the first few thousand feet the water content is reduced to approximately 30 percent (by volume), of which 20 to 25 percent is interlayer water and 5 to 10 percent pore water. These values for interlayer water would seem to be high. Muds containing nonswelling clays contain pore water, and it would seem reasonable that swelling clays contain pore water and that it would be a significant proportion of the total water, particularly when the expanded layers comprise less than 50 percent of the bulk rock.

For the Chevron samples, the difference in weight loss at room temperature and 110° C is approximately 2.5 percent. This value remains constant with depth. When the samples were dried at 300° C, the weight loss (from room

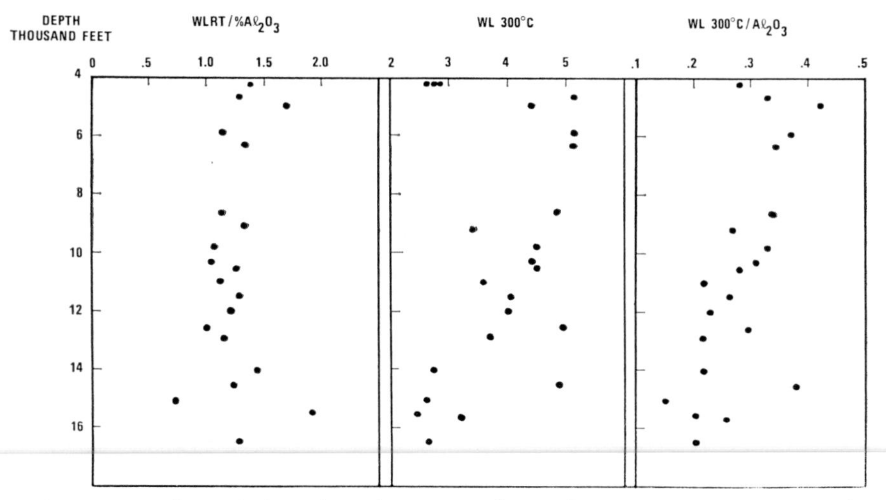

Figure 23. Left graph shows how the percent of water lost at room temperature varies as a function of percent Al_2O_3. Middle graph gives the amount of water lost between room temperature and 300° C (percent of 300° C weight). Right graph shows how percent of water lost at 300° C varies as a function of percent Al_2O_3.

temperature weight) ranged from 2.75 to 5.1 percent (Fig. 23). These latter values are presumably a relative measure of the amount of interlayer water; it is difficult to determine how they relate to the subsurface water values.

The amount of water lost at 300° C decreases with depth; most of the water is lost below 10,000 ft. The relatively high clay content of the samples from 10,500 ft to 13,000 ft, in part, accounts for the large water loss in the deeper samples. The amount of water lost at 300° C per percent of Al_2O_3 decreases fairly regularly from 4000 ft to 16,450 ft. Samples below 10,530 ft appear to lose less water per unit of depth than the shallow samples. This observation is in agreement with the X-ray data. Part of the difference in slope is due to the fact that the shallow and deep samples have slightly different clay mineral/ Al_2O_3 ratios. There may be an abrupt, but small, decrease in interlayer water across the permeability barrier, but the data are not reliable enough to prove it.

Across the permeability barrier (9800 ft to 11,000 ft) the pore water increases by 5 percent and the interlayer water decreases by 0.5 percent. The contrast is slightly less on a percent of Al_2O_3 basis. Through this same interval, the Al_2O_3 values increase by 2 percent, equivalent to approximately 10 percent clay minerals. The expanded clays would have to lose six layers of water to account for this increase in pore water. It is unlikely that the clays could contain this much interlayer water at depths where a good permeability seal could form.

One-tenth gram of water per gram of montmorillonite gives one water layer between two montmorillonite sheets (Hendricks and others, 1940). The difference in weight loss between room temperature and 300° C is from 0.044 to 0.051 grams per gram of solid in the shallow samples and from 0.025 to 0.032 in the deepest samples. These data indicate the expanded layers of the shallow samples (~50 percent) contain one water layer, and those of the deeper samples (~10 percent) contain two water layers, and are in agreement with the exchange data which indicates Na is the dominant cation in the shallow samples, and Ca, in the deeper samples.

If the expandable layers contain one layer of water, this water amounts to 3 to 5 percent by weight of the dry solid as compared to 18 to 25 percent pore water (this latter value is probably low as some water was likely lost before the samples were analyzed). If the expandable layers contained two water layers (in the ground), this would amount to 6 to 10 percent of the weight of the dry sediment. Interlayer water is certainly a significant component, but by no means does it overshadow the pore water.

A plot of percent water lost at room temperature that varies as a function of percent Al_2O_3 (Fig. 23), shows a decrease trend to ~7000 ft; deeper samples have a relatively constant ratio. There does appear to be a relative increase across the permeability barrier, but the anomalous values appear to be the low values for the barrier samples (9800 ft and 10,300 ft) rather than the high values for the samples beneath the barrier.

If the interlayer water is removed by simple dehydration, then fresh water is supplied to the pores. If the expandable layers are converted to illite and chlorite, then a cation-rich solution may be released to pores. The cation concentration in the last few layers of interlayer water is much larger than that of seawater, and these cations should dominate the pore water chemistry. The calculated molarity for two water layers ranges from 7 to 10 (Jackson, 1963, p. 37). The molarity of seawater is 0.7.

If some of the K used to make illitic layers is obtained from K-feldspar, and the Al and Fe used to make chloritic layers are obtained from K-feldspar, kaolinite and free oxides, then interlayer cations are replaced by ions from the solid phase, and there should be a net increase in the pore water population. A further increase in the cation (H^+) concentration should be caused by the formation of hydroxy-Al and hydroxy-Fe polymers in the interlayer position. This increase in cation concentration would require an increase in anion concentration. This, in part, may account for the high HCO_3 values found in the interstitial waters.

There is no major clay mineral change across the permeability barrier. The decrease in expandable layers appears to be a gradual change with a decrease in rate with an increase in depth. The nature and rate of change appears to be the same as in the other wells in the area that do not have abnormal fluid pressures.

The preceding data suggest that the abnormally high water pressures are

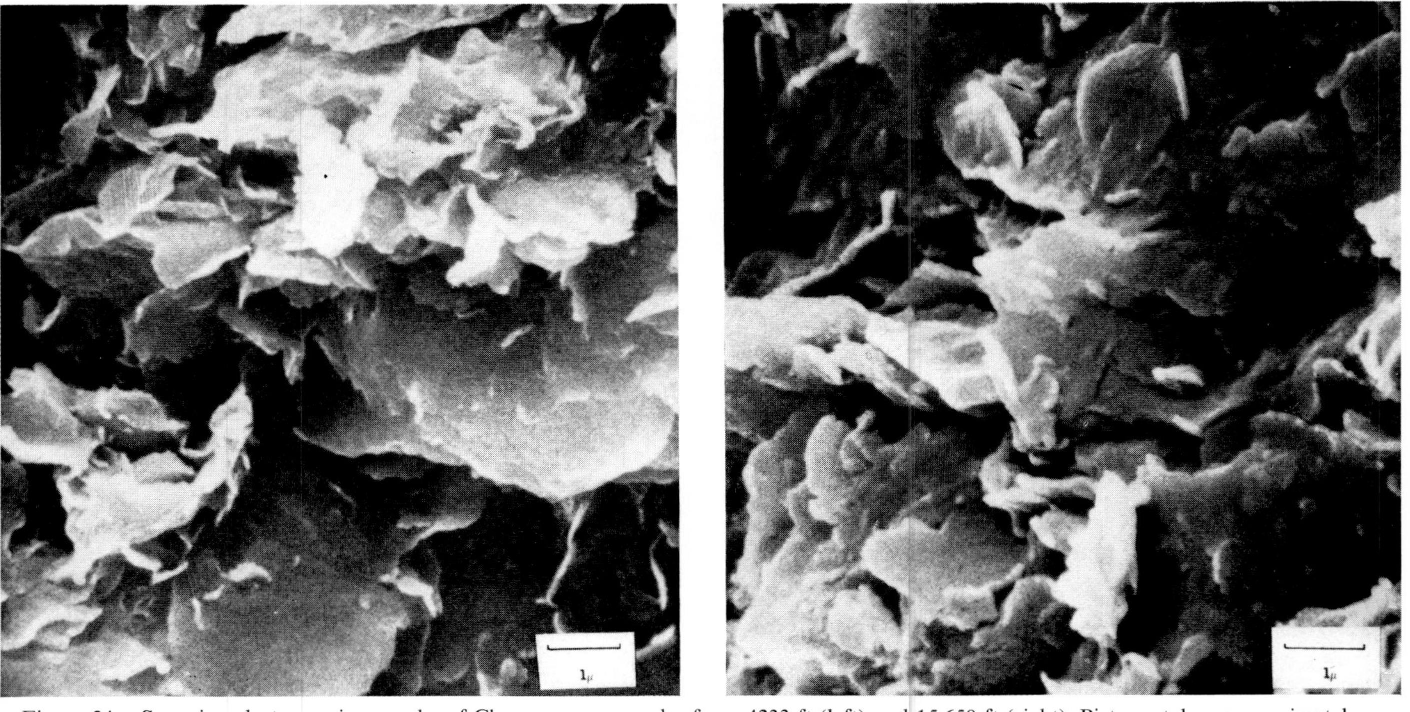

Figure 24. Scanning electron micrographs of Chevron core samples from 4233 ft (left) and 15,650 ft (right). Pictures taken approximately perpendicular to bedding.

due to the formation of a physical barrier due to compaction caused by the rapid dewatering of the top part of a thick section of water-rich montmorillonitic mud. Upward migration of ions, water, and organics is hindered. Ca precipitates as calcite and thereby increases the effectiveness of the barrier. Ions and water continue to migrate upward to the barrier. Some of the more mobile ions, Na and Cl, pass through the barrier more easily than H_2O. Interlayer water is added to the pore water throughout the section, but it does not appear to be a major contributor in the deeper samples.

Although a barrier relatively impermeable to water appears to exist between 10,230 and 10,500 ft, it need not prevent ion diffusion. White (1965) has reviewed the ideas on fine-grained sediments behaving as semipermeable membranes, permitting the selective escape of water and concentrating dissolved components, and found the evidence convincing. He further suggested that as the salinity of pore fluids increased some cations could be removed with the escaping waters. Such a mechanism probably accounts for the distribution of ions above approximately 9800 ft in the Chevron well.

Ramberg (1952) maintained that ions can readily diffuse through rocks where the permeability is so low that the fluid phase cannot. This is particularly true in clay-rich sediments. The clay minerals have high-surface areas, high-surface charge, and well-developed surface-water layers. Lai and Mortland (1960) and others have shown that ions can diffuse rapidly through muds without water movement. Lai and Mortland suggested that self-diffusion of cations in a clay-water system occurred by cations jumping from the exchange site on one particle to a site on another particle. Activation energy decreased with increasing clay content, presumably because exchange sites are closer together. In the Miocene muds both a pressure and geothermal gradient provide the necessary driving energy.

Scanning electron microphotographs were made of fractured surfaces perpendicular to the bedding for samples from 4233 ft and 15,650 ft. There is some suggestion of more orientation and closer flake packing in the deep sample (Fig. 24), but the differences are not striking.

Conclusions and Speculations

Weathering has caused Al to be concentrated in the hydrolyzates, primarily in the shales. Once the burial cycle has started and temperatures begin to increase, the Al-rich clay suite begins a slow and persistent evolution toward a greater than 300° to 400° C equilibrium mineral suite of muscovite, chlorite, and quartz.

During the first stage, inherited exchangeable K migrates to the highest charged layers, and inherited interlayer hydroxyl-Al and hydroxyl-Fe becomes better organized and more stable. Most of the calcite is dissolved in this early stage, and the Ca is flushed from the mud. With further temperature increase, the lattice charge of some layers is increased by the reduction of Fe and by the incorporation of some interlayer Al into the tetrahedral layer. K from K-feldspar and perhaps biotite is adsorbed by these newly created high-charge layers increasing the amount of 10 Å material. Some of the Al from the K-feldspar may move into other interlayers, increasing the number of dioctahedral chlorite layers. This mechanism presumably accounts for the relatively minor amount of K-feldspar found in most shales and low-rank metamorphic rocks. The resulting mixed-layer clay is composed of montmorillonite, 10 Å "illite," and dioctahedral chlorite layers. These layers form below 100° C. With increasing temperature, the regularity of the interlayering increases with little additional change in the proportion of layers. The interlayer material (K, Al, Fe) is apparently relatively mobile and moves to produce a layer arrangement (13 Å phase) that has a high degree of stability. This phase consists of approximately equal parts of the three components.

At a temperature near 150° C, kaolinite starts to decompose, and the Al as hydroxy-Al moves into the interlayer position increasing the proportion of dioctahedral chlorite layers. At this stage some of the chlorite layers form packets with a sufficient number of layers to diffract as the discrete mineral chlorite. Some additional Al may move into the tetrahedral layer at this stage and some packets of 10 Å layers form (the K derived from K-feldspar). Thus, the amount of discrete 10 Å "illite" and dioctahedral chlorite has increased slightly, but the majority of the clay consists of a mixed-layer illite-chlorite with a lesser amount of montmorillonite.

Up to this stage of diagenesis, the total chemistry is changed only slightly, if at all. Little further progress toward the muscovite-chlorite-quartz assemblage will occur unless additional ions (mainly K, Mg, Fe) are supplied to the system to balance the high Al content.

With increasing temperature ($\sim$200° C), either due to deeper burial or increase in heat flow rates, upward-migrating K, Mg, and Fe, derived from the underlying sediments, become sufficiently abundant that the remaining expanded layers are lost, and some discrete 10 Å illite and trioctahedral chlorite formed; however, much of the "illite" at this stage still contains appreciable proportion of dioctahedral chlorite, and the "chlorite" contains some 10 Å layers. This is the typical clay mineral suite found in Paleozoic and Precambrian shales. Detrital mica and chlorite (trioctahedral), derived largely from metamorphic rocks, is also a sizable component of these shales.

Complete unmixing of the mixed-layered minerals requires temperatures on the order of 400° C or a relative high degree of metamorphism (phengite $\rightarrow$ muscovite + chlorite). Metamorphism commonly stops at this stage, and the rocks reach the surface of the crust where weathering and Al-enrichment start, and a new cycle begins.

If this postulated mechanism is reasonably correct, it implies that the pre-middle Carboniferous shales were, in general, exposed to higher temperatures than the majority of younger shales and muds. This can be due to deeper burial or higher heat flow rates, or both. In the geosynclinal areas both increased rate of flow and depth are factors, and conversion tends more toward completion.

The relatively abrupt change in the chemistry and mineralogy of shales which occurs in the middle Carboniferous suggests that heat flow rates may have decreased relatively rapidly at this time. The global distribution pattern of this chemical and mineral change should provide information about the history of the Earth's crust. Chemical and mineral variations in shales may be extremely subtle indications of variations in temperature and rates of heat flow.

The exchangeable cations and interstitial ions also undergo diagenetic changes as the temperature increases and the solid phase changes.

The most notable features of the exchange cations, in addition to a decrease in abundance with depth, is the early loss of Mg to form dolomite in the mud and chlorite in the sands and the dominance of Na in the shallow samples and Ca in the deep samples.

In the surface samples the exchange suite is Mg $>$ Na; Ca $>$ K; at a depth of 4000 ft, the sequence is Na $>$ Ca $>$ Mg $>$ K; at 17,000 ft, the suite is Ca $>$ Mg $>$ K $>$ Na. At a relatively shallow depth the exchange cations and exchange sites become more abundant than the cations in the interstitial water. Ion mobility becomes more important than charge, and Na replaces Mg as the major exchange cation. With increasing depth and temperature, interlayer sites tend to become occupied (and unavailable for cation exchange) by K, Al, and Fe, as illitic and chloritic layers are formed. The weakly bonded Na is preferentially removed, and Ca is retained, along with Mg, in the exchange position.

K apparently moves from solid phase (K-feldspar) to solid phase (10 Å layers) without causing a major increase in exchangeable K, though the K content of the water phase is consistently larger than that of seawater.

The preferential Ca retention in the exchange position could explain the increase in the Ca/Cl ratio, with age, of brines from sandstones and limestones (Chave, 1960, p. 365). During advance diagenesis, when the last expanded layers are made nonexpandable, Ca may be mobilized to help form the calcite commonly found in veins permeating low-rank metamorphic sediments. The Mg presumably goes to form chlorite.

As the mud section is buried, both interstitial and interlayer water are forced from the muds. The Na content decreases during shallow burial; about 20 percent is adsorbed on the clays in exchange for Mg and about 40 percent migrates from the mud along with Cl. Cl continues to be lost from the mud with increasing burial. Because of the high rate of deposition, SO_4 is not reduced and is concentrated by selective filtration. HCO_3 is formed from calcite and organic material and becomes the dominant anion.

As the water-rich montmorillonite bathyal muds are dewatered, a physical permeability barrier is formed at the top of the mud section and is caused by the relatively rapid loss of water through the overlying sandy shelf sediments. With increased temperature, adsorbed organic material is released from the clay minerals and, along with the ions and water, migrates upward. Due to a decrease in temperature and pressure and an increase in Na and SO_4, Ca precipitates as calcite increasing the effectiveness of the barrier. The upward migration of water is blocked, and abnormally high water content and pressures develop. The release of interlayer water to the pores is of minor importance below 10,000 ft (80° C).

Na migrates upward, systematically decreasing the amount in the deeper muds; Na reaches a maximum in the permeability barrier. Na apparently diffuses through the barrier more easily than H_2O. Cl diffuses through the barrier more easily than SO_4 and HCO_3.

K and Mg in the intersititial water increases systematically in the deeper samples. This is presumably due to the increased temperature and the increased solubility of the silicate minerals. Some of the K, Mg, and Fe may have been released under deeper, higher temperature conditions and migrated upward from below. This increase in ion concentration may be a shallow indication of a deep-seated granitization.

References Cited

Alexiades, C. A.; and Jackson, M. L. Quantitative clay mineralogical analysis of soils and sediments: Clays Clay Miner., 14th Conf., Pergamon Press, p. 35–52, N.Y., 1966.

Bates, T. F. Investigation of the micaceous minerals in slate: Amer. Mineral., Vol. 32, p. 625–636, 1947.

Burst, J. F., Jr. Post diagenetic clay mineral-environmental relationships in the Gulf Coast Eocene: Clays Clay Miner., Proc. Natl. Conf., Vol. 6, p. 327–341, 1958.

Burst, J. F., Jr. Diagenesis of Gulf Coast clayey sediments and its possible relation to petroleum migration: Amer. Ass. Petrol. Geol., Bull., Vol. 53, p. 73–93, 1969.

Chave, K. E. Evidence of history of sea water from chemistry of deeper subsurface waters of ancient basins: Amer. Ass. Petrol. Geol., Bull., Vol. 44, p. 357-370, 1970.

Clark, F. W. The data of geochemistry: U.S. Geol. Surv., Bull., No. 770, 841 p., 1924.

Davidtz, J. C.; and Low, P. F. Effect of lattice configuration on swelling of montmorillonites (Abstr.): Clays Clay Miner., 17th Ann. Conf., p. 11, 1968.

Dickey, P. A.; Shriram, C. R.; and Paine, W. R. Abnormal pressures in deep wells in southwestern Louisiana: Science (AAAS), Vol. 160, No. 3828, p. 609–615, 1968.

Dickinson, George. Reservoir pressures in Gulf Coast Louisiana: Amer. Ass. Petrol. Geol., Bull., Vol. 37, p. 410–432, 1953.

Dunoyer De Segonzac, G. Les argiles du Cretace Superior dans le bassin de Douala (Cameroun): Problemes de diagenese: Bull. serv. carte geol. Als. Lorr., Vol. 17, p. 287-310, 1964.

Dunoyer De Segonzac, G.; and Kubler, B. Sur la cristallinte de l'illite dans la diagenese et l'anchimetamorphisme: Sedimentology, Vol. 10, p. 137-143, 1968.

Engelhardt, W. V.; and Gaida, K. H. Concentration changes of pore solutions during the compaction of clay sediments: J. Sediment. Petrology, Vol. 33, p. 919–930, 1963.

Ernst, G. Significance of phengitic micas in low grade schists: Amer. Mineral., Vol. 48, p. 1357–1373, 1963.

Flawn, P. T.; Goldstein, A., Jr.; King, P. B.; and Weaver, C. E. The Ouachita System: Univ. Texas Pub., No. 6120, p. 401, 1961.

Fuchtbauer, H.; and Goldschmidt, H. Boebachtungen zur tonmineral-diagenese: 1st Int. Conf. Clay Proc., Stockholm, p. 99–111, 1963.

Glenn, R. C.; and Nash, V. E. Weathering relationships between gibbsite, kaolinite, chlorite, and expansible layer silicates in selected soils from the lower Miss. Coastal Plain: Clays Clay Miner., Pergamon Press, Vol. 12, p. 529–548, N.Y., 1964.

Glenn, R. C.; Jackson, M. L.; Hole, F. D.; and Lee, G. B. Chemical weathering of layer silicates clays in loss-derived tama silt loam of southwestern Wisconsin: Clays Clay Miner., Pergamon Press, Vol. 8, p. 63–83, N.Y., 1960.

Goldberg, E. D.; and Arrhenius, G. O. S. Chemistry of Pacific pelagic sediments: Geochim. et Cosmochim. Acta, Vol. 13, p. 53–212, 1958.

Green-Kelly, R. Dehydration of the montmorillonite minerals: Mineral. Mag., Vol. 30, p. 604–615, 1955.

Grim, R. E.; and Johns, W. D. Clay mineral investigation of sediments in the northern Gulf of Mexico: Clays Clay Miner., Nat. Acad. Sci., Nat. Res. Council, Publ., Vol. 327, p. 81–108, 1954.

Hanshaw, B. B. Cation-exchange constants for clays from electronical measurements: Clays Clay Miner., Pergamon Press, Vol. 12, 397–421, N.Y., 1964.

Hendricks, S. B.; Nelson, R. A.; and Alexander, L. T. Hydration mechanisms of the clay mineral montmorillonite saturated with various cations: J. Amer. Chem. Soc., Vol. 62, p. 1457–1464, 1940.

Holmes, A. Principles of Physical Geology: Ronald Press, p. 1288, 1965.

Irving, E. Paleomagnetism of some Carboniferous rocks from New South Wales and its relation to geologic events: J. Geophys. Res., Vol. 71, p. 6025–6051, 1966.

Jackson, M. L. Interlayering of expansible layer silicates in soils by chemical weathering: Clays Clay Miner., Pergamon Press, Vol. 13, p. 29–46, N.Y., 1963.

Jones, P. H. Hydrodynamics of geopressure in the northern Gulf of Mexico Basin: Amer. Inst. Min. Eng. Mtg., SPE Pap., No. 2207, p. 12, 1968.

Kerr, P. F.; and Barrington, J. Clays of deep shale zones, Caillore Island, Louisiana: Amer. Ass. Petrol. Geol., Bull., Vol. 45, p. 1697–1712, 1961.

Khitarov, N. I.; and Pugin, V. A. Behavior of montmorillonites under elevated temperatures and pressures: Geochem. Int., Vol. 3, No. 4, p. 621–626, 1966.

Klipp, R. W.; and Barney, J. E., II. Determination of sulfur traces in naphthas by lamp combustion and spectrophotometry: Anal. Chem., Vol. 31, p. 596–597, 1959.

Kossovskaya, A. G. Uber die spezifischen epigenetischen umwandlungen terrigener gesteine in Tafelland-und geosynklinal-gebieten: Doklady Akad. Nank SSSR., Vol. 130 (1), p. 176–179, 1960.

Kossovskaya, A. G.; and Shutov, V. D. Epigenetic zones in the Mesozoic and Upper Paleozoic clastic sedimentary rocks of the Western Verkhoyansk region: Doklady Akad. Nank SSSR, Vol. 103, No. 6, 1955.

Kubler, B. Les angiles, indicateurs de metamorphisme: Rev. Inst. France Petrole, Vol. 19, p. 1093–1112, 1964.

Kuznetsova, V. A. The propagation of sulfate-reducing bacteria in the oil deposits of the Kaybyshev Oblast in connection with the salt composition of the stratal waters: Mikrobiol., Vol. 29, No. 3, 1960.

Lai, T. M.; and Mortland, M. M. Self-diffusion of exchangable cations in bentonite: Clays Clay Miner., Pergamon Press, Vol. 9., p. 229–247, N.Y., 1960.

Lambert, R. St. J. The mineralogy and metamorphism of the Maine schists of the Morar and Konydart district of Inverness-Shire: Roy. Soc. Edinburgh, Trans., Vol. 63, p. 53–588, 1959.

Lee, W. H. K. Terrestrial heat flow: Geophy. Monogr. Ser., No. 8., Am. Geophys. Union, Nat. Acad. Sci., Nat. Res. Council, Pub. No. 1288, 1965.

Lee, W. H. K.; and Uyeda, Seiya. Review of heat flow data in terrestrial heat flow: Am. Geophys. Union Geophys. Monogr., Vol. 8, p. 100, 1965.

Levorsen, A. L. Geology of Petroleum: W. H. Freeman and Co., p. 723, San Francisco, 1967.

MacEwan, D.M.C.; Ruiz Amil, A.; and Brown, G. Interstratified clay minerals, the X-ray identification and crystal structures of clay minerals: Mineral. Soc. London, p. 393–445, 1961.

Marshall, C. E. Multifunctional ionization as illustrated by the Clay minerals: Clays Clay Miner., Nat. Acad. Sci., Nat. Res. Council, Pub. No. 327, p. 364–385, 1954.

Martin, R. T. Adsorbed water on clay: a review: Clays Clay Miner., Pergamon Press, Vol. 9, p. 28–70, N.Y., 1960.

Mitchell, B. D.; and Mackenzie, R. C. Free iron oxide removal from clays: Soil Sci., Vol. 77, p. 173, 1954.

Muffler, L. J.; and White, D. E. Active metamorphism of Upper Cenozoic sediments in the Salton Sea geothermal field and the Salton Trough, Southeastern California: Geol. Soc. Amer., Bull., Vol. 80, p. 157–182, 1969.

Muller, G. Diagenesis in argillaceous sediments, *in* Diagenesis in Sediments: Elsevier Pub. Co., 551 p., N.Y., 1967.

Myers, R. L.; Van Siclen, C. D. Dynamic phenomena of sediment compaction in Metagorda County, Texas: Gulf Coast Ass. Geol. Soc., Trans., Vol. 14, p. 241–252, 1964.

Nanz, R. G., Jr. Chemical composition of Precambrian slates with notes on the geochemical evolution of lutites: J. Geol., Vol. 61, p. 51–64, 1953.

Overton, H. L.; and Zanier, A. M. An osmotic model for gas and overpressured formations: (in press) 1971.

Philippi, G. T. On the depth, time and mechanism of petroleum generation: Geochem. et Cosmochim. Acta, Vol. 29, p. 1021–1050, 1965.

Poldervaart, A. Chemistry of the earth's crust, *in* Crust of Earth, A. Poldervaart (ed): Geol. Soc. Amer., Sp. Paper 62, p. 119–144, 1955.

Porter, J. R. Bacterial chemistry and physiology: John Wiley and Co., 1073 p., N.Y., 1946.

Post, D. F.; and White, J. L. Clay mineralogy and mica-vermiculite layer charge density distribution in the Switzerland soils of Indiana: Soil Sci. Soc. Amer., Proc., Vol. 31, p. 419–424, 1967.

Powers, M. C. Adjustment of clays to chemical change and the concept of the equivalence level: Clays Clay Miner., Proc., Nat. Conf., Vol. 6, p. 309–326, 1958.

Powers, M. C. Fluid release mechanisms in compacting marine mudrocks and their importance in oil exploration: Amer. Ass. Petrol. Geol., Bull., Vol. 51, p. 1240–1254, 1967.

Radoslovich, E. W.; and Norrish, K. The cell dimensions and symmetry of layer-lattice silicates I.—Some structural considerations: Amer. Mineral., Vol. 47, p. 599–616, 1962.

Rainwater, F. H.; and Thatcher, L. L. Methods for collection and analysis of water samples: U.S. Geol. Surv., Water-Supply Pap. 1454, 301 p., 1960.

Ramberg, H. The origin of metamorphic and metasomatic rocks: Univ. Chicago Press, p. 317, Chicago, 1952.

Raman, K. V.; and Jackson, M. L. Layer charge relations in clay minerals of micaceous soils and sediments: Clays Clay Miner., Pergamon Press, N.Y., 14th Conf., p. 53-68, 1966.

Rex, R. W. Authigenic kaolinite and mica as evidence for phase equilibria at low temperatures: Clays Clay Miner., Vol. 13, p. 95–104, 1964.

Reynolds, R. C., Jr. Interstratified clay systems: Calculations of the total one-dimensional diffraction: Amer. Mineral., Vol. 52, p. 661–662, 1967.

Rich, C. I. Hydroxy interlayers in expansible layer silicates: Clays Clay Miner., Vol. 16, p. 15–30, 1968.

Rich, C. I.; and Obenshain, S. S. Chemical and clay mineral properties of a red-yellow Podzolic soil derived from muscovite schist: Soil Sci. Soc. Amer., Proc., Vol. 19, p. 334–341, 1955.

Rittenberg, S. C.; Emery, K. O.; and Orr, W. L. Regeneration of nutrients in sediments of marine basins: Deep-Sea Res., Vol. 3, p. 23-45, 1955.

Ross, C. S. Clays and soils in relation to geologic processes: J. Washington Acad. Sci., Vol. 33, p. 225, 1943.

Scotford, D. M. Petrology of the Cincinnatian Series shales and environmental implications: Geol. Soc. Amer., Bull., Vol. 76, p. 193–222, 1965.

Shaw, D. M. Geochemistry of pelitic rocks. Part III, Major elements and general geochemistry: Geol. Soc. Amer., Bull., Vol. 67. p. 919–934, 1956.

Shaw, D. B.; and Weaver, C. E. The mineralogical composition of shales: J. Sediment. Petrology, Vol. 35, p. 213-222, 1965.

Siever, R.; Beck, K. C.; and Berner, R. A. Composition of interstitial waters of modern sediments: J. Geol., Vol. 73, p. 39–73, 1965.

Steiner, A. Clay minerals in hydrothermally altered rocks at Wairakei, New Zealand, (Abstr.): Clays Clay Miner., 16th Conf., p. 31–33, 1967.

Stone, R. L., and Rowland, R. A. DTA of kaolinite and montmorillonite under water vapor pressures up to six atmospheres: Clays Clay Miner., 3rd Nat. Conf., Nat. Acad. Sci., Nat. Res. Council Pub. 395, p. 103-116, 1955.

Swain, F. M. Geochemistry of humus: Organic Geochem. Monogr., The MacMillan Co., p. 87–147, N.Y., 1963.

Tageyeva, N. V. Interstitial waters of Arctic Ocean sediments: Doklady Akad. Nank SSSR, vol. 163, p. 207-209, 1965.

Tettenhorst, R. Cation migration in montmorillonites: Amer. Mineral., Vol. 47, p. 769–773, 1962.

Turekian, K. K.; and Wedepohl, K. H. Distribution of the elements in some major units of the earth's crust: Geol. Soc. Amer., Bull., Vol. 72, p. 175–192, 1961.

Van der Plas, L. Petrology of the Northern Adula Region, Switzerland: Leidse Geologische Medelingen, Vol. 24, p. 418–599, 1959.

Van Olphen, H. Collapse of potassium montmorillonite clays upon heating—'potassium fixation': Clays Clay Miner., 14th Nat. Conf., p. 393-406, 1966.

Velde, B. Mixed-layer mineral associations in muscovite-celadonite and muscovite: Clays Clay Miner., 13th Nat. Conf., p. 29–32, 1964.

Vinogradov, A. P.; and Renov, A. B. Evolution of the chemical composition of clays of the Russian platform: Geochemistry, Vol. 2, p. 123–138, 1956.

Warshaw, C. M. Experimental studies of illite: Clays Clay Miner., 7th Conf., Pergamon Press, p. 303–316, N.Y., 1959.

Weaver, C. E. The effects and geological significance of potassium 'fixation' by expandable clay minerals derived from muscovite, biotite, chlorite, and volcanic material: Amer. Mineral., Vol. 43, p. 839–861, 1958a.

Weaver, C. E. The clay petrology of sediments: Clays Clay Miner., Proc., 6th Nat. Conf., p. 154–187, 1958b.

Weaver, C. E. Possible uses of clay minerals in the search for oil: Amer. Ass. Petrol. Geol., Bull., Vol. 44, p. 1505–1518, 1960.

Weaver, C. E. Potassium content of illite: Science, Vol. 147, No. 3658, p. 603–605, 1965.

Weaver, C. E. The significance of clay minerals in sediments, *in* Fundamental aspects of petroleum geochemistry, Nagy, B. and Colombo, U., (eds.), p. 37–75, 1967a.

Weaver, C. E. Potassium, Illite and the Ocean: Geochim. et Cosmochim. Acta, Vol. 31, p. 2181–2196, 1967b.

Weaver, C. E. Relation of composition to structure of dioctahedral 2:1 clay minerals: Clays Clay Miner., Vol. 16, p. 51–61, 1968.

Weaver, C. E. Chemistry of clay minerals: U.S. Geol. Surv. (in press), 1971.

Weaver, C. E.; and Wampler, J. M. K, Ar, Illite Burial: Geol. Soc. Amer., Bull., Vol. 81, No. 11, p. 3423–3430, 1970.

White, D. E. Magmatic, connate and metamorphic waters: Geol. Soc. Amer., Bull., Vol. 68, p. 1659–1682, 1957.

White, D. E. Saline waters of sedimentary rocks: Amer. Ass. Petrol. Geol., Mem. 4, p. 342–366, 1965.

Wickman, F. E. The "total" amount of sediments and the composition of the "average igneous rock": Geochimica et Cosmochim. Acta, Vol. 5, p. 97–110, 1954.

Wilson, A. D. The micro-determination of ferrous iron in silicate minerals by a volumetric and a colorimetric method: Analyst, Vol. 85, p. 823–827, 1960.

Winkler, H. G. F. Cas T-P der diagenese und niedrigemperierten metamorphose aufgrand von mineralreactionen: Beitr. Z. Mineral. U. Metrogr., Vol. 10, p. 70–93, 1964.

Winkler, H. G. F. Petrogenesis of metamorphic rocks: Springer-Verlag, N.Y., 1967.
Yaalon, D. H. Downward movement and distribution of anions in soils profiles with limited wetting: Exper. Ped., p. 157–164, 1965.

MANUSCRIPT RECEIVED BY THE SOCIETY AUGUST 10, 1970

Appendix: Semidisplacive Mechanism for Diagenetic Alteration of Montmorillonite Layers to Illite Layers

CHARLES O. POLLARD, JR.

ABSTRACT

Illite/montmorillonite (in bulk and in mixed-layer clays) increases during diagenesis, indicating that montmorillonite is altered to illite. Altering average sedimentary montmorillonite layers to average diagenetic illite layers principally involves increasing tetrahedral Al from almost nil to ~20 percent. Regarding possible mechanisms: (1) Al probably cannot displace Si and occupy the same tetrahedron; (2) high-energy barriers undoubtedly exist at submetamorphic temperatures to a mechanism requiring disintegration; (3) during dehydration of montmorillonite, some exchangeable cations (with radius $\leq Na^{+}$) apparently enter through the "ditrigonal" holes in basal oxygen planes into expanded tetrahedral sites (one hydroxyl ion, three basal oxygen ions); and (4) the expanded tetrahedral sites contract to normal 4 CN packing as tetrahedral rotation increases to the theoretical maximum (30°). The proposed mechanism involves: (a) migration of interlayer Al through ditrigonal holes; (b) increased tetrahedral rotation (attraction of basal oxygens toward the Al ions in the expanded tetrahedra); (c) creation of permanent Al tetrahedra at 30° rotation; (d) expulsion of an equal number of Si to the interlayer volume; (e) additional Si migration along easy paths to new (illite) tetrahedra; (f) hydrogen ion migration to nearest neighbors within the oxygen-hydroxyl planes; and (g) continued rotation to the illite configuration. Processes (c) through (f) occur simultaneously. This mechanism is sterically possible at each structural level. Inherited illite OH-OH angles within individual layers depend on the spatial relations between the sets of tetrahedra vacated in top and bottom sheets, and on the original montmorillonite OH-

OH angles. Poor stacking order in montmorillonite predetermines the 1 Md illite polytype, as has been observed in early diagenetic illites. The (Al, Si) tetrahedra are ordered in the product, illite, within each single-layer domain.

INTRODUCTION

Several studies (Burst, 1959; Powers, 1959; Weaver, 1959) have shown that illite layers replace montmorillonite layers in the clay mineral fraction of rocks undergoing diagenesis. Interlayer hydroxide sheets may create the appearance that more than the actual number of such alterations occur; nevertheless, the replacement affects a substantial volume of material in those columns of marine sediments over approximately 10,000 ft thick as demonstrated by Weaver and Beck (p. 19). The principal change that must be effected to alter the average sedimentary montmorillonite layer to the average diagenetic illite layer is an increase in the amount of tetrahedral Al. Weaver (1959) has argued that the need for tetrahedral Al may be apparent, as "most analyses of montmorillonite are of commercial grade, relatively pure deposits and are probably not typical"; however, his suggestion that most expanded clays carried into a marine environment are beidellitic (high Al), and that potassium, heat, and pressure are the only requisites to form mixed layer illite-montmorillonite, is unproven. If the starting layers are normal montmorillonite, the substitution of Al in tetrahedra increases from essentially nil to approximately 20 percent in the resulting illite layers. If the starting material is more beidellitic, some increase in tetrahedral Al is necessary for alteration to illitic layers, as the tetrahedral substitution in most expanded clays carried into the marine environment is probably intermediate between 0 percent and the maximum of 12 percent given by Weaver (1959).

In a discussion of the montmorillonite-to-illite alteration, Powers (1959) has proposed that Al that has been displaced by Mg from the octahedral sheet may replace Si in the tetrahedral sheets. Without definite steric models, it is difficult to judge whether or not Mg, Al, and Si are likely to be mobile enough within the clay layer for such replacements to occur, short of disintegration of the structure (at least to the separation of silica tetrahedra from the octahedral sheet). In discussing this alteration, Towe (1962), Weaver (1959) and Powers (1959) have implied that the basic three-sheet clay layer remains intact throughout, which seems reasonable; high-energy barriers presumably would preclude a reaction path involving disintegration (especially at submetamorphic temperatures). Moreover, direct displacement of Al from oxygen octahedra by Mg, or of Si from oxygen tetrahedra by Al, as proposed by Powers, is unlikely, even in detached polyhedra.

On the other hand, if the three-sheet layers retain integrity during diagenesis, the mobility of ions (in particular, Al and Si) into and out of the tetrahedral sheets needs attention. Replacement (exchange) of cations within the interlayer volume presents no difficulty; in fact, mobility of some ions through this volume suggests it as a convenient pipeline and storage space. Al and Si, the ions whose mobility is under scrutiny, are certainly not highly mobile in the interlayer volume, but Al (at least) can migrate therein well enough to disperse gibbsite crystallites throughout, even under weathering conditions (Shen and Rich, 1962; Jackson, 1963). Tettenhorst (1962) has presented strong, infrared evidence that some small cations enter the ditrigonal holes (Radoslovich, 1960) in the tetrahedral sheets adjacent to the interlayer volume during dehydration and collapse of montmorillonite. Tettenhorst did not study the effects of dehydration on a montmorillonite that had been Al-exchanged (Sawhney, 1968). The possibility that Al can also enter the holes in basal oxygen planes is interesting with respect to the mechanism for Al substitution into montmorillonite tetrahedra during diagenesis. (Such a possibility is now being explored experimentally.) A potential mechanism for montmorillonite-layer diagenesis, involving this Al-migration, is proposed in detail below. The interlayer volume serves both as the supply route for Al to the adjacent tetrahedral sheet and as the escape route for displaced Si.

The site occupied by the smaller cations (Li^+, Mg^{2+}, Ni^{2+}, Cu^{2+}, Zn^{2+}, Mn^{2+}, and possibly Na^+ and Ca^{2+}) in the collapsed montmorillonites of Tettenhorst (1962) is an expanded tetrahedron (with three basal oxygens and an OH loosely "co-ordinated" around the previously exchangeable cation). The entrance holes into the tetrahedral sheet are the "ditrigonal holes" (Radoslovich, 1960) in the basal oxygen plane. Al may migrate along the same paths in Al-enriched systems, but more drastic steric readjustments are necessary to bring about the montmorillonite-illite alteration than simple entrance by Al to the level of the tetrahedral sheet through the ditrigonal holes. The presence in the "loose tetrahedra" of such highly charged cations as Al probably initiates the steric readjustments.

The tightness of packing in the tetrahedral-sheet "loose tetrahedra" depends (under normal conditions) on the octahedral (di- or tri-) character as well as the tetrahedral rotation, tetrahedral tilt, tetrahedral-sheet thickness, and other deviations from the idealized phyllosilicate structure (Bailey and others, 1967). These loose sites contract toward more normal 4 CN packing in response to increased tetrahedral tilting and rotation. (Actually, tetrahedral rotation and tilting are interdependent, but much of the following discussion is in terms of rotation alone.) A wide range of tetrahedral rotations is observed in dioctahedral phyllosilicates, up to 21° in margarite (Takéuchi, 1965). The maximum theoretical rotation is 30°, at which value the packing in the "loose tetrahedron" is equivalent to that in the original Si tetrahedra. Whereas the "loose tetrahedra" become more similar to one of the sites preferred by Al (4 CN with respect to oxygen) with increasing tetrahedral rotation, the Al ions in the "loose tetrahedra" will produce a drive to increase tetrahedral rotation toward 30° by attracting the residual charges on basal oxygen ions. If the concentration of Al in the "loose tetrahedra" becomes high enough during sediment diagenesis, it is expected that the drive to increase tetrahedral rotation will provide the activation energy, as it were, to alter the montmorillonite layers to illite layers, with accompanying expulsion of Si to the interlayer volume and, thence, outside the clay.

THE PROPOSED MECHANISM

The basic process in the proposed mechanism is easily followed in Figure 1, which shows the reorganization expected in a tetrahedral sheet. A detailed analysis is required of the relations between such a tetrahedral sheet and the other tetrahedral sheet of the same layer, at each stage of the proposed alteration. The changes that occur in the octahedral sheet depend on the changes in the two tetrahedral sheets and on the nature of the initial (montmorillonite) octahedral sheet; therefore, a classification of the three possibilities for articulation of lower and upper O-OH planes within a single layer [1], illustrated in Figure 2, is desirable for ease of discussion. The model assumed for the montmorillonite single layer is essentially based on that presented by Hofmann and others (1933) and MacEwan (1961), with the additional assumption that the dioctahedral Al ions and vacancies within a sheet are ordered, as has been observed in all dioctahedral phyllosilicates reported to have been well-crystallized enough for such determination (Bailey and others, 1967). Dioctahedral-ionic ordering restricts, by causing distortions of the structure, the possible layers to those with OH-OH angles (in basal projection) of any one of three 120° rotations (Smith and Yoder, 1956; Radoslovich, 1959, 1960; Takéuchi, 1965), here labeled layer types I, II, and III. In terms of the alteration being proposed, it is necessary to examine the results of starting with each of these three possible OH-OH angles within an individual montmorillonite

[1] The layer type here labeled type I is the only one of the three reported in the modern structural refinements of dioctahedral 2:1 phyllosilicates. The possible existence of dioctahedral 2:1 layer types II and III apparently has not been considered since Radoslovich (1959, 1960) pointed out that the idealized phyllosilicate structure based on hexagonal arrays of ions was an inadequate model. Bloss and others (1963) have considered explicitly the possibility of the three layer types in the trioctahedral mica, fluorphlogopite, where the three are degenerate (indistinguishable) in the perfect (untwinned, undomained), isolated single layer. Such is not the case for the dioctahedral layer types I, II, and III; layer type I is energetically different from layer types II and III, which are mirror-related.

When I first traced the various possibilities for alteration of a montmorillonite (type I) layer, I was forced to consider the additional two possible layer types (II and III) in illite. The existence of layer types II and III as well as type I is obviously pertinent to the problem of compositional control on the occurrence of mica polytypes; if only one of three possible intralayer stackings has been considered, no complete model of mica polytypism could have been developed (Radoslovich, 1960; Güven and Burnham, 1967; Güven, 1968; Franzini, 1969). Development of such a model is being attempted, taking into account the three-intralayer stacking possibilities.

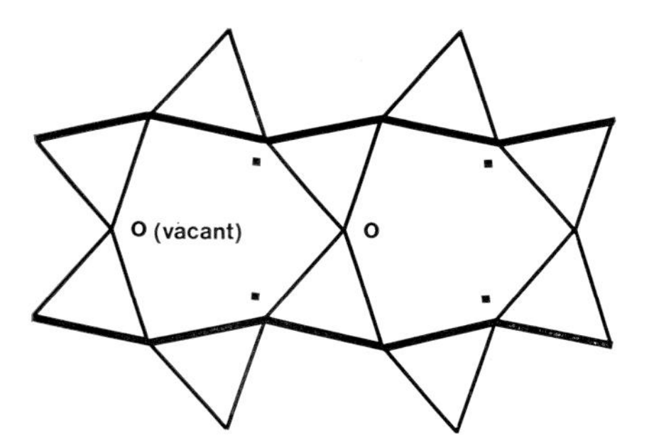

A. Projection on (001) of the basal faces of tetrahedra in one sheet and of occupied and vacant octahedral sites for a montmorillonite layer. Tetrahedral rotation is approximately 12°, and tetrahedra are tilted approximately 5.5° about the bolder edge. This is the bottom tetrahedral sheet, so that the basal oxygen in each tetrahedron that is not on the bolder edge is slightly elevated.

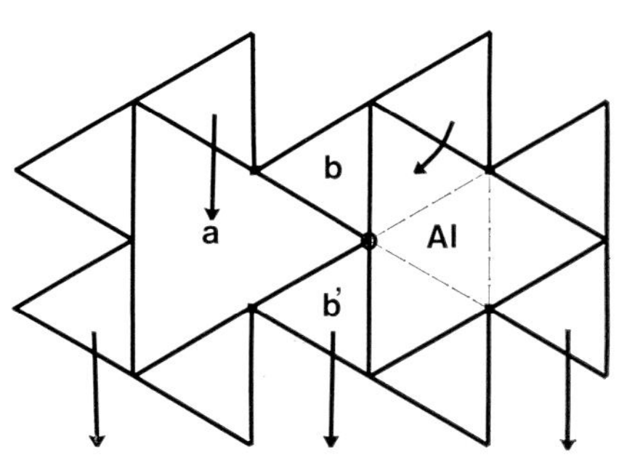

B. Aluminum has entered some of the sites *a*, causing full rotation of the tetrahedra (30°) and creating the possibility that silicon can be expelled either from the tetrahedral sites *b* or *b′*. The *b′* expulsion is schematically depicted here. The expelled Si can migrate along easy paths either to empty sites *a* (straight arrows) or to the interlayer volume (curved arrow, if neighboring *a* sites are occupied by Al).

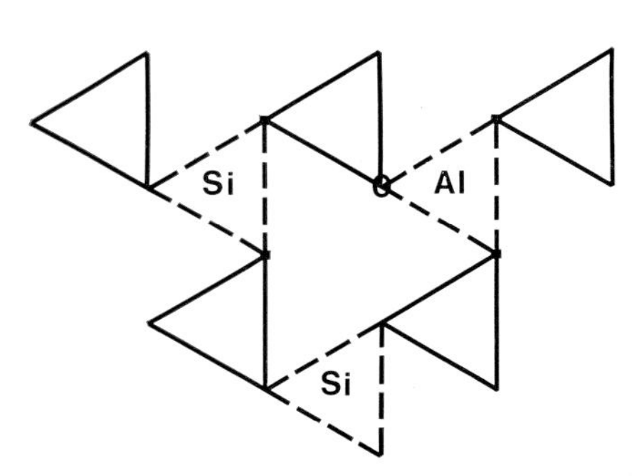

C. Previous *b′* sites are depicted as empty and previous *a* sites are depicted as occupied either by Al or by Si that has migrated from previous *b′* sites. These new (Al, Si) tetrahedra are outlined by dashed lines. The previous *b* tetrahedra are unchanged, still occupied by Si, and outlined in solid lines. Relaxation from extreme 30° rotation has not occurred yet at this stage.

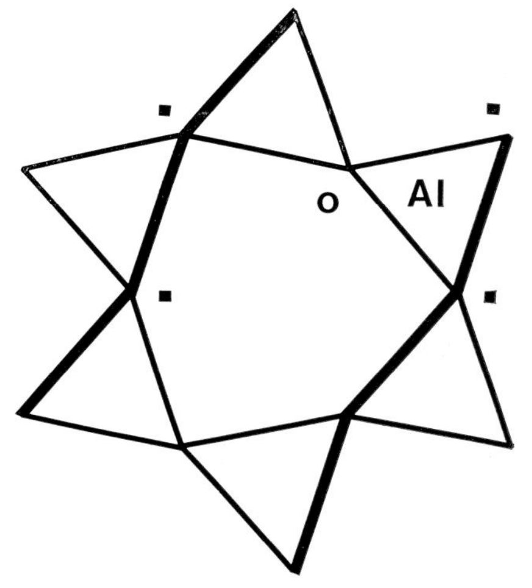

D. Continued tetrahedral rotation and retilting of tetrahedra yield the bottom tetrahedral sheet of the alteration product, where the corrugations in the basal oxygen plane trend at 120° from those in the original tetrahedral sheet. The distribution of Al over the tetrahedral sites in the alteration product is ordered, and the ratio Al/Si in the (Al, Si) tetrahedra varies from 0 to 1/3 (as depicted) to infinity, depending on the occupancy of the original sites *a* by Al when increased tetrahedral rotation was initiated.

Figure 1. Proposed alteration of a single tetrahedral sheet.

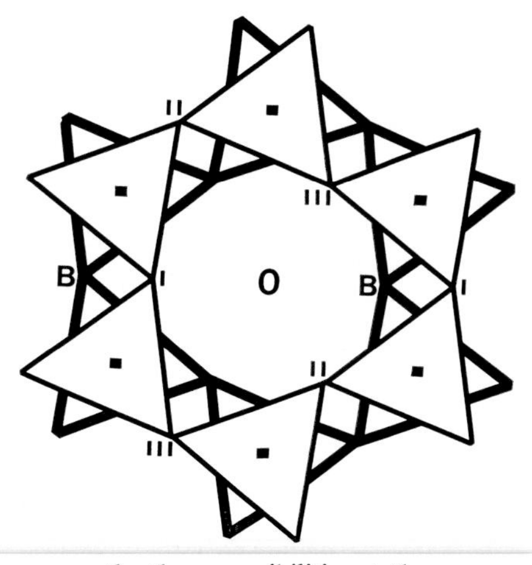

Figure 2. Classification of layer types, based on articulation of O-OH planes. Apical oxygens and hydroxyls in the bottom O-OH plane occupy the corners of the bolder triangles, which are the lower faces of distorted octahedra around occupied octahedral sites. *B* labels the OH location in the bottom O-OH plane for all possibilities; that is, as reference, the basal projection of the line through the vacant octahedra and OH in the bottom O-OH plane is kept horizontal here. The top O-OH plane can mesh with the bottom O-OH plane in any of three ways, but the top hydroxyls and apical oxygens project on the corners of the less bold triangles for all three possibilities. Considering only the octahedral sheet, the differences among the three possible layer types relate to the identity of the OH group among the three possibilities at the corners of each upper octahedron face. Layer types I, II, and III are defined as shown here, depending on the location of the OH in the upper O–OH plane relative to OH in the lower O-OH plane.

layer. No average or idealized model for individual layers would be satisfactory. Figure 3 shows the top and bottom tetrahedral sheets both projected on (001), for each of the layer types. All three layer types are geometrically feasible. There is no *a priori* need to have parallel corrugations in the upper and lower basal-oxygen planes of the same layer, as found in layer type I.

The crystal structures that should serve as the bases for Figures 1a, 1d, 2, and 3 are naturally those of montmorillonite and illite. Barring available structures for these minerals, the pyrophyllite structure would be most logical as a substitute for montmorillonite and the muscovite structure should represent illite. (A phengite structure with lower total charge and lower octahedral charge than that analyzed by Güven, 1968, would be appropriate for illite, but is apparently unavailable.) An adequately precise structure determination is not available for pyrophyllite either (Rayner and Brown, 1966), so the muscovite structure is reluctantly used for both illite and montmorillonite. As the structure is not exactly suited for the clays anyway, Figures 1a, 1d, 2, and 3 are schematic to the extent that bond lengths and angles shown are slightly more regular than in published muscovite structures (Radoslovich, 1960; Güven and Burnham, 1967), for ease in drafting. The tetrahedral rotation assumed for these figures is approximately 12° and the tetrahedral tilt is about 5 5°. Neighboring O-O separations are approximately 2.68 Å in Figures 1, 3, and 4. Neighboring O-O or O-OH separations are approximately 2.87 Å in Figure 2.

In agreement with Franzini (1969), the tetrahedral rotations in these dioctahedral clays are represented as "positive-positive"; that is, all layers

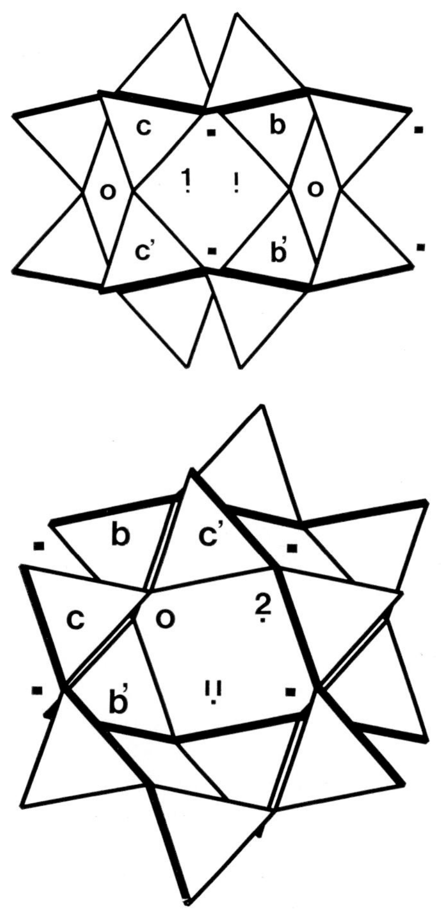

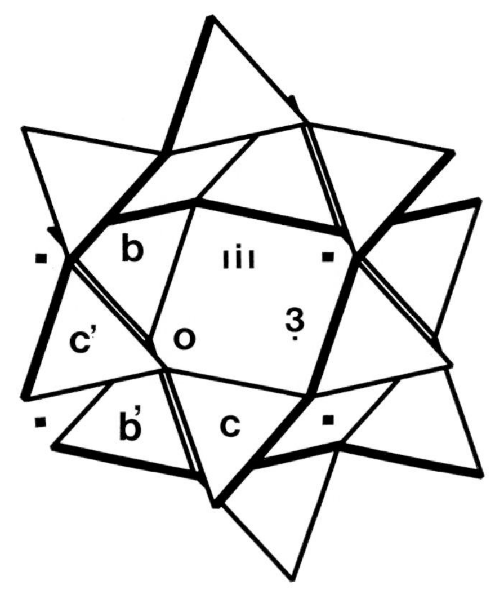

Figure 3. Projection onto (001) of basal faces of upper and lower tetrahedra, upper and lower hydroxyls, occupied and vacant octahedral sites. Arabic numerals designate hydroxyls in the lower O-OH plane; Roman numerals designate hydroxyls in the upper O-OH plane; *b* and *b′* label the two unique tetrahedral sites in lower tetrahedral sheets; *c* and *c′* label the two tetrahedral sites in upper tetrahedral sheets. Bolder tetrahedral edges represent axes of tetrahedral tilting in both sheets. The corrugations are depressed in the lower tetrahedral sheet and elevated in the upper tetrahedral sheet. Note that the trend of tetrahedral corrugations in each sheet parallels the basal projection of the line from the vacant octahedral site to the nearest hydroxyl in the same sheet. Figure 2 can be used also to relate the corrugations in lower tetrahedral sheets to corrugations in upper tetrahedral sheets, for each layer type. A. Layer type I (top). B. Layer type II. C. Layer type III.

depicted are A layers (*see* Franzini, 1969), which signifies that the tetrahedral rotation in both tetrahedral sheets is such that the basal oxygens are closer (in the basal projection) to the nearest octahedral sites of the same layer than they would be in the idealized structure where basal oxygens are arrayed in hexagonal planar patterns.

Tetrahedral Sheets

Arbitrarily, a type I individual montmorillonite layer (Figs. 2 and 3a) is chosen to illustrate in detail the proposed alteration mechanism. The analysis applies similarly to layer-types II and III. The alteration of the bottom tetrahedral sheet is illustrated in Figure 1, and is the same for all the possibilities. Al ions entering the sites marked *a* through the ditrigonal holes are expected to cause an increased tetrahedral rotation because of the attraction toward these ions of the three nearest basal oxygen ions. Such rotation would make the *a* site more nearly like a normal tetrahedron (including the nearest OH ion in the fourfold co-ordination). If the tetrahedral rotation were to increase to 30°, the *a* site would be a close-packed tetrahedron similar to the original Si tetrahedra [2]. From that value, the tetrahedral rotation could change only by restretching the Al-O bonds in the *a*-site tetrahedra (if the rotation decreases), or by stretching and breaking the Si-O bonds in either *b*-site or *b′*-site tetrahedra (if the rotation increases).

As stated, it may seem unlikely that the obvious path to lower energy lies toward stretched Si-O bonds rather than toward stretched Al-O bonds. Indeed, if too few *a* sites were occupied by Al ions, the attraction would not be widespread enough to cause increased tetrahedral rotation near the critical 30° value. Once a critical concentration of Al ions developed at the *a* sites, the layer could be visualized as an activated complex, such that if thermal vibrations ever caused rotation past 30°, the Si ions would be expelled from *b* or *b′* sites irreversibly. As shown in Figure 1d, the end result is a tetrahedral sheet with alternate tetrahedra occupied by (Al, Si), the Al/Si ratio depending on the concentration of exotic Al ions in the *a* site "voids" during the activated complex stage. The critical concentration of Al ions at *a* sites in montmorillonite may well be around one-third Al/*a*-site, which would yield an altered tetrahedral sheet with one-sixth of the tetrahedra occupied by Al ions, as in the average illite. One-half Al/*a*-site would yield a micalike tetrahedral sheet; one Al/*a*-site would yield the tetrahedral sheet of a brittle mica.

At the same time that Si ions are expelled from *b* or *b′* sites in the bottom tetrahedral sheet, Si ions are also expelled from analogous sites in the top tetrahedral sheet. *c* and *c′* are the top tetrahedra analogous to *b* and *b′* bottom tetrahedra; for the spatial relationships among *b*, *b′*, *c*, and *c′*, consult Figure 3. No control is obvious that would predict a preference for expulsion from

---

[2] The Al tetrahedron is actually somewhat larger than the Si tetrahedron, the quantitative differences depending on the local structural environments (Smith and Bailey, 1963). A different empirical relation describes the dependence of mean tetrahedral cation-to-oxygen distance on tetrahedral Al content for each silicate structural type. It is difficult, therefore, to select a probable value for Al-O distance in the proposed intermediate structure (Fig. 1b), which is certainly not a classical sheet silicate structure. Assuming that all tetrahedra are regular, and that Si-O distances and Al-O distances can take on values from 1.61Å to 1.63Å and from 1.74Å to 1.80Å, respectively (as suggested *by* Fig. 4 *of* Smith and Bailey, 1963), the extreme tetrahedral rotation that would be exhibited in the proposed intermediate structure would be between 26° and 27¾°, rather than 30°. Because the *a*-site tetrahedron is not necessarily a regular tetrahedron (even if the *b* and *b′* sites are), the significance of these extreme rotational angles is uncertain.

either of the b or b' sites in the bottom sheet to correlate with one or the other analogous expulsions in the top tetrahedral sheet; therefore, the assumption is made that pure chance determines which of the four possible combinations (b-c, b-c', b'-c, or b'-c') of expulsions obtains.

Because, for the moment, the discussion is limited to the alteration of type I layers, the relationship between top and bottom tetrahedral sheets in the beginning clay layer is as shown in Figure 3a. If at the same time Si ions are expelled from b' sites in the bottom tetrahedral sheet, c' site expulsions also occur in the top sheet, then the relationship between top and bottom sheets in the alteration product would be as shown in Figure 3c. These new relative orientations are arrived at by applying the same analysis represented in Figures 1a through 1d to each of the original tetrahedral sheets, as relatively oriented in Figure 3a. If the combination of expulsions is from b and c sites, beginning with a type I layer, the resultant is as shown in Figure 3b. If the combination of expulsions is b-c' or b'-c, another layer develops with the relationship between top and bottom tetrahedral sheets as shown in Figure 3a, but the whole layer is rotated by $\pm$ 120°. After considering the effects of these alterations on the octahedral sheets, the various possibilities can be more easily classified (*in* Table 1).

Octahedral Sheet

In response to entrance by Al into sites a (Fig. 1), tetrahedral rotation is expected to increase, causing geometric readjustments in the octahedral sheet as well. One potential geometric readjustment is illustrated by the change from one of the configurations observed in montmorillonite (Fig. 2, where I designates the OH in the top O-OH plane) to the assumed configuration when the tetrahedral rotation is extreme at 30° (Fig. 4, where 1 and I designate the hydroxyl ions in the bottom and top O-OH planes, respectively). The hexagonal array of O and OH ions in the O-OH planes at extreme tetrahedral rotation is inescapable if it is assumed that octahedral O-O and O-OH separations are at least as large as basal tetrahedral O-O separations; that is, tetrahedral tilt is nil when tetrahedral rotation is 30°. The first portion of the proposed alteration in the octahedral sheet is thus geometric readjustment; an exchange reaction follows.

The changes in the octahedral sheet are represented in Figure 4 as occurring concomitantly with an alteration involving expulsions of Si from b-type bottom tetrahedra and from c-type top tetrahedra. Again, a type I beginning layer is chosen as an example. As the a-sites are altered to regular (Al, Si) tetrahedra and the b-sites are vacated by Si ions, a concomitant exchange of the proton is required between the previous OH groups coordinated around the a-sites and the previous apical oxygen ions of the b-sites. A similar exchange occurs in the O-OH plane shared with the top tetrahedral sheet, between the previous OH-ions and the previous apical oxygen ions of the *c*-sites. The result is shown in Figure 4, where 2 and II designate the new hydroxyl ions in the bottom and top O-OH planes, respectively. Continued rotation of the tetrahedra would lead to a configuration similar to that in Figure 2, where the OH in the top O-OH plane is labelled II. The *b-c* alteration of an original montmorillonite type I layer thus results in an illite type II layer. Similar analysis shows that a *b′-c′* alteration of a type I layer yields a type III layer (as also was demonstrated by consideration of the tetrahedral sheets). In fact, Table 1 shows the possibilities for the various combinations of beginning layer type, sites of bottom tetrahedral expulsion, and sites of top tetrahedral expulsion.

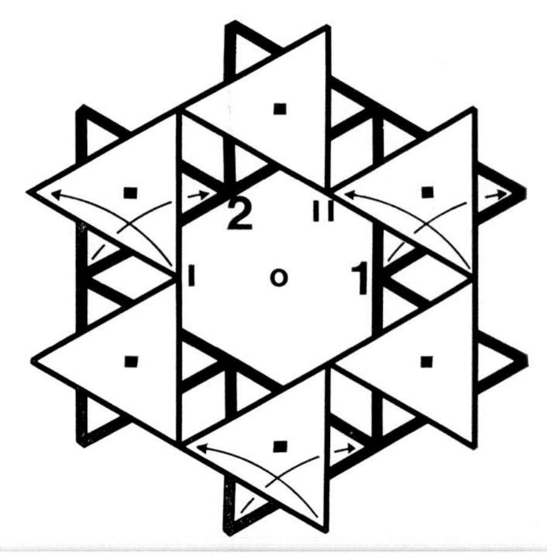

Figure 4. Proton exchanges among oxygens and hydroxyls in the *b-c* alteration of a type I layer to a type II layer. At this stage tetrahedral rotation is extreme at 30° and tetrahedral tilting is 0°. Octahedral O and OH separations are at least as large as basal tetrahedral O-O separations. Arabic numerals 1 and 2 designate initial and final hydroxyls in the lower O-OH plane, respectively. Roman numerals I and II designate initial and final hydroxyls in the upper O-OH plane, respectively. Arrows represent proposed proton exchanges

POLYTYPES

Because of the "turbostratic" or disordered (at least in the "long-term" sense) arrangement among montmorillonite layers (MacEwan, 1961), a realistic model for the montmorillonite-illite alteration cannot be extended to a larger portion of the clay structure than the individual layer. It has been demonstrated above that only certain restricted illite individual layer types (regarding OH-OH angles) can develop from an individual montmorillonite layer depending on the vagaries of the initial alteration steps; however, it is impossible to account for the attainment of any more layer-to-layer ordering in the illite than existed in the original montmorillonite. Thus Table 1 is pertinent only to alterations of individual layers. This aspect of the proposed mechanism is in full accord with observations on low-grade illites, which are known to progress from 1 Md to 2 Ml polytypes with metamorphic grade (Maxwell and Hower, 1967). The present mechanism includes no device for ordering the 1 Md polytype of illite once it is attained.

TABLE 1. LAYER TYPES FOR INITIAL AND FINAL LAYERS, WITH THE SITES OF SI EXPULSION IN TOP AND BOTTOM TETRAHEDRAL SHEETS

Initial layer type	Si expulsion * Bottom tetrahedral sheet	Top tetrahedral sheet	Final layer type
I	*b*	*c*	II
I	*b*	*c'*	I
I	*b'*	*c*	I
I	*b'*	*c'*	III
II	*b*	*c*	III
II	*b*	*c'*	II
II	*b'*	*c*	II
II	*b'*	*c'*	I
III	*b*	*c*	I
III	*b*	*c'*	III
III	*b'*	*c*	III
III	*b'*	*c'*	II

* If *b* is the site of Si expulsion in the bottom sheet of the initial layer, then the tetrahedra marked *b'* in the appropriate final layer (Fig. 3) will have increased Al content; that is, the *b'* tetrahedra are (Al, Si) tetrahedra. A *b'* Si expulsion leads to a final layer in which the b tetrahedra are (Al, Si) tetrahedra. Likewise, a *c* Si expulsion in the top sheet of the initial layer leads to a final layer in which the *c'* designates (Al, Si) tetrahedra. A *c'* Si expulsion leads to a final layer in which *c* designates (Al, Si) tetrahedra.

DOMAINS

The proposed mechanism might affect different portions of the same phyllosilicate layer differently, as Si may be displaced from different combinations of tetrahedra (*b* or *b'* below and *c* or *c'* above) along the layer. Whether the boundaries marking such faults are called twin composition surfaces (Bloss and others, 1963) or domain boundaries (Gatineau, 1964), the (Al, Si) tetrahedra are ordered relative to the (Si) tetrahedra and the layer type (I, II, or III) is the same within each boundary. Whereas no control is apparent to select either top tetrahedral expulsion (*c* or *c'*) to accompany a given bottom (*b* or *b'*) expulsion, the domain (twin) boundaries probably involve faults in only one of the two tetrahedral sheets. Faults in the product layer also can arise from inherited faults within the original montmorillonite layer, that is, domains (twins) in the illite layer may arise from changes laterally within a layer of any of the controlling factors in Table 1.

The expectation that the montmorillonite-illite alteration would give rise to illites within which the layers are in domains (twins), possibly of small dimensions, suggests that layer-to-layer order would be unattainable for a significant (regarding X-ray powder diffraction) volume of the product illite. The product would assume the aspect of a 1 Md clay mineral if substantial numbers of domain (twin) boundaries developed during the alteration, even if the initial clay particle had been an ideal, ordered crystal (which is, of course, not true of montmorillonite). The possibility then arises that the 1Md

$\rightarrow 2M_1$ trend during diagenesis of illite (Maxwell and Hower, 1967) involves annealing of domains (twins) so that energetically preferred combinations of layers replace energetically detrimental combinations of layers (Radoslovich, 1960). The details of the $1Md \rightarrow 2M_1$ transformation are outside the subject of this article, but are no less interesting than those of the montmorillonite-illite alteration. In fact, Gatineau (1964) has reported chainlike domains, each with one of three different patterns of ordering in (Si, Al) tetrahedra running parallel to the XY plane of the structure, in a thorough study of diffracted intensity throughout the reciprocal space associated with a muscovite crystal.

ALUMINUM CONCENTRATION

Two aspects of the concentration of Al are important: (1) the Al concentration in the interstitial water, and (2) the critical Al concentration in the sites *a*.

Aluminum in the intersititial water apparently becomes readily available early during diagenesis of fine-grained sediments because of breakdown of potassium feldspar, at least to concentrations high enough to facilitate formation of some interlayer Al $(OH)_3$; later during diagenesis the Al is provided also by breakdown of kaolinite. There is strong indication that some of the released Al exists as the dissolved species in the interlayer volume of clays (particularly at low OH/Al ratios), rather than all the Al organizing into interlayer Al $(OH)_3$ (Sawhney, 1968; Shen and Rich, 1962). The dissolved Al is then eligible for migration into the *a* sites.

A critical concentration of Al in the *a* sites is probably required to cause extreme tetrahedral rotation. This critical concentration may be lower than the one-third Al/*a* site that would result in an illite layer from one operation of the proposed mechanism. If the critical concentration is less than one-third, the mechanism would be required to operate more than once during diagenesis, such that tetrahedral Al would increase with each operation. In this sense, the montmorillonite-to-illite alteration may represent a segment of the larger diagenetic-metamorphic alteration from montmorillonite to mica, the total process being constructed of smaller semidisplacive steps. The ordered distribution of (Al, Si) tetrahedra (within each domain) after one such semidisplacive alteration should give rise to adjustments in bond lengths and angles so that each layer type (I, II, or III) will have a preferred alteration scheme (*b*-*c*, *b*-*c*′), which effectively could increase tetrahedral ordering, together with each increase in tetrahedral Al. Interlayer K sites also could become more distorted in a regular sense with increased tetrahedral ordering, potentially providing an increasing interlayer control on the association of OH-OH angles in neighboring layers and an increasing steric preference for one of the polytypes (presumably $2M_1$).

Of course, K-O (basal) bonds do not remain intact during each semidisplacive step in the alteration, as the K site moves during each step from a position below the *a* site in basal projection to a position below those sites originating as *b* or *b′* sites. In the less highly charged clays, at least, such disregard for the interlayer bonding is in accord with the cation-exchange and swelling characteristics, in spite of the importance Franzini (1969) and Radoslovich (1960) have placed on the interlayer bonding in the more highly charged micas. Thus the increasing layer charge may place a limit on the number of semidisplacive steps that can operate on a given phyllosilicate layer, by increasing interlayer cations that can bar further alteration through exertion of relatively permanent control over the relative orientations of adjacent layers.

Thus an alternative to the domain-annealing explanation for the 1Md → 2 M_1 transformation (in illite) given above is available, but in this latter possibility the 1Md polytype should have associated with it a lower content of tetrahedral Al than should the $2M_1$ polytype. The tendency toward tetrahedral ordering and toward a certain polytype described here need not result finally in the same ordering and polytype found in muscovite by Radoslovich (1960), Gatineau (1964), or Güven and Burnham (1967); none of their samples were of burial diagenetic-metamorphic origin. The phengite analyzed by Güven (1968) is representative of a totally different metamorphic genesis than the more "normal" burial diagenesis and metamorphism of concern here (Weaver and Beck, p. 20–21; Maxwell and Hower, 1967); hence, even the ordering and polytype found in that sample are not pertinent.

SUMMARY AND CONCLUSION

During the diagenetic alteration of montmorillonite layers to illite layers, Al may indirectly displace Si from the tetrahedral sheet, after having entered through the ditrigonal holes in the basal oxygen plane. (The steps involved in the process are succinctly and sequentially listed in the abstract.) The proposed mechanism is appealing because Al is not put into direct competition with Si for the tetrahedral sites, and because it is topotactic (Glasser and others, 1963) or semidisplacive, which avoids obviously high energy barriers. A precise description of a manner in which montmorillonite can acquire tetrahedral Al and interlayer cations as it is altered to illite, as well as release Si for silica cement, is thus provided.

ACKNOWLEDGMENTS

I wish to thank the donors of the Petroleum Research Fund, administered by the American Chemical Society, for support of this research. Discussions with Charles E. Weaver were essential, and he was kind enough to read and criticize the manuscript. Criticisms by S. W. Bailey and Lin D. Pollard also are appreciated.

REFERENCES CITED

Bailey, S. W.; Brindley, G. W.; Keller, W. D.; Wones, D. R.; and Smith, J. V. Layer Silicates: Amer. Geol. Inst. Short Course Lecture Notes, 1967.

Bloss, F. D.; Gibbs, G. V.; and Cummings, D. Polymorphism and twinning in synthetic fluorophlogopite: J. Geol., Vol. 71, p. 537–547, 1963.

Burst, J. F., Jr. Postdiagenetic clay mineral environmental relationships in the Gulf Coast Eocene: Clays Clay Miner. Proc. 6th Natl. Conf., p. 327–341, 1959.

Franzini, M. The A and B mica layers and the crystal structure of sheet silicates: Contr. Mineral. Petrol., Vol. 21, p. 203–224, 1969.

Gatineau, L. Structure réele de la muscovite. Répartition des substitutions isomorphes: Soc. Franç. Minér. Crist. Bull., Vol. 87, p. 321–355, 1964.

Glasser, L. S. D.; Glasser, F. P.; and Taylor, H. F. W. The role of oriented transformations in mineralogy: Mineral. Soc. Amer., Spec. Pap. 1, p. 200–203, 1963.

Güven, N. The crystal structures of 2 M_1 phengite and 2 M_1 muscovite: Carnegie Inst. Wash., Yearbook 66, p. 487–492, 1968.

Güven, N.; and Burnham, C. W. The crystal structure of 3T muscovite: Z. Kristallogr., Vol. 125, p. 163–183, 1967.

Hofmann, V.; Endell, K.; and Wilm, D. Strucktur und Quellung von Montmorillonit: Z. Kristallogr., Vol. 86, p. 340–348, 1933.

Jackson, M. L. Aluminum bonding in soils: A unifying principle in soil science: Soil Sci. Soc. Amer. Proc., Vol. 27, No. 1, p. 1–10, 1963.

MacEwan, D. M. C. Montmorillonite minerals, in the X-ray identification and crystal structures of clay minerals (Ed. by G. Brown): Mineral. Soc., p. 143–207, London, 1961.

Maxwell, D. T.; and Hower, J. High-grade diagenesis and low-grade metamorphism of illite in the Precambrian Belt series: Amer. Mineral., Vol. 52, p. 843–857, 1967.

Powers, M. C. Adjustment of clays to chemical change and the concept of the equivalence level: Clays Clay Miner., Proc., 6th Nat. Conf., p. 327–341, 1959.

Radoslovich, E. W. Structural control of polymorphism in micas: Nature, Vol. 183, p. 253, 1959.

Radoslovich, E. W. The structure of muscovite, $KAl_2 (Si_3Al) O_{10} (OH)_2$: Acta Crystallogr., Vol. 13, p. 919–932, 1960.

Rayner, J. H.; and Brown, G. Structure of pyrophyllite: Clays Clay Miner., Proc., 13th Nat. Conf., p. 73–84, 1966.

Sawhney, B. L. Aluminum interlayers in layer silicates; effect of OH/Al ratio of Al solution, time of reaction, and type of structure: Clays Clay Miner., Vol. 16, p. 157–164, 1968.

Shen, M. J.; and Rich, C. I. Aluminum fixation in montmorillonite: Soil Sci. Soc. Amer. Proc., Vol. 26, p. 33–36, 1962.

Smith, J. V.; and Bailey, S. W. Second review of Al-O and Si-O tetrahedral distances: Acta Crystallog., Vol. 16, p. 801–811, 1963.

Smith, J. V.; and Yoder, H. S. Experimental and theoretical studies of the mica polymorphs: Min. Mag., Vol. 31, p. 209–235, 1956.

Takéuchi, Y. Structures of brittle micas: Clays Clay Miner., Proc., 13th Nat. Conf., p. 1–25, 1965.

Tettenhorst, R. Cation migration in montmorillonites: Amer. Mineral., Vol. 47, p. 769-772, 1962.

Towe, K. M. Clay mineral diagenesis as a possible source of silica cement in sedimentary rocks: J. Sediment. Petrology, Vol. 32, p. 26–28, 1962.

Weaver, C E. The clay petrology of sediments: Clays Clay Miner., Proc., 6th Nat. Conf., p. 154-187, 1959.

MANUSCRIPT RECEIVED BY THE SOCIETY NOVEMBER 16, 1970

Index